9th International Workshop on High-pT Physics at LHC

Journal of Physics: Conference Series Volume 589

Grenoble, France
24-28 September 2013

ISBN: 978-1-5108-0078-6
ISSN: 1742-6588

Printed from e-media with permission by:

Curran Associates, Inc.
57 Morehouse Lane
Red Hook, NY 12571

Some format issues inherent in the e-media version may also appear in this print version.

Printed by Curran Associates, Inc. (2015)

For permission requests, please contact the Institute of Physics
at the address below.

Institute of Physics
Dirac House, Temple Back
Bristol BS1 6BE UK

Phone: 44 1 17 929 7481
Fax: 44 1 17 920 0979

techtracking@iop.org

Additional copies of this publication are available from:

Curran Associates, Inc.
57 Morehouse Lane
Red Hook, NY 12571 USA
Phone: 845-758-0400
Fax: 845-758-2634
Email: curran@proceedings.com
Web: www.proceedings.com

Table of contents

Volume 589

Accepted papers received: 26 January 2015
Published online: 10 March 2015

Papers

Journal of Physics: Conference Series **589** (2015) 011001 doi:10.1088/1742-6596/589/1/011001

Preface

This volume contains selected papers presented at the *9th International Workshop on High-p_T at LHC*. The workshop was held from 24-28 September 2013 at LPSC Grenoble, France.

The first workshop of this series was organized in 2006 in Trento, Italy. The purpose of the 9th workshop was to offer an opportunity for experimentalists and theoreticians to get together and discuss experimental results from the first heavy-ion program at LHC, lessons learnt from RHIC and theoretical developments. Focus was also given to high p_T physics in pA collisions, especially at the LHC energies. Emphasis was given on discussion over a large scope: high-p_T physics, jets, photons, correlations, hard scattering and hard probes phenomena.

The main topics of the workshop were:

- Nuclear modifications of the parton distribution functions
- High p_T jet production in pp, pA and AA
- High p_T parton propagation in matter
- Nuclear modifications of the fragmentation functions
- Correlations with leading particles
- Direct photon and heavy flavor tagging

About 50 participants coming from 12 countries participated in the workshop. 6 invited talks and 31 oral contributions were presented at the conference.

Journal of Physics: Conference Series **589** (2015) 011002 doi:10.1088/1742-6596/589/1/011002

Peer review statement

All papers published in this volume of *Journal of Physics: Conference Series* have been peer reviewed through processes administered by the proceedings Editors. Reviews were conducted by expert referees to the professional and scientific standards expected of a proceedings journal published by IOP Publishing.

Journal of Physics: Conference Series **589** (2015) 012001 doi:10.1088/1742-6596/589/1/012001

Monte Carlo Studies of Identified Two-particle Correlations in p-p and Pb-Pb Collisions

Gyula Bencédi[1], **Gergely Gábor Barnaföldi**[1], **Levente Molnár**[2]

[1] Wigner Research Centre for Physics of the Hungarian Academy of Sciences, Budapest, Hungary

[2] Institut Pluridisciplinaire Hubert Curien, Strasbourg, France

E-mail: bencedi.gyula@wigner.mta.hu

Abstract. Azimuthal particle correlations have been extensively studied in the past at various collider energies in p-p, p-A, and A-A collisions. Hadron-correlation measurements in heavy-ion collisions have mainly focused on studies of collective (flow) effects at low-p_T and parton energy loss via jet quenching in the high-p_T regime. This was usually done without event-by-event particle identification. In this paper, we present two-particle correlations with identified trigger hadrons and identified associated hadrons at mid-rapidity in Monte Carlo generated events. The primary purpose of this study was to investigate the effect of quantum number conservation and the flavour balance during parton fragmentation and hadronization. The simulated p-p events were generated with PYTHIA 6.4 with the Perugia-0 tune at $\sqrt{s} = 7$ TeV. HIJING was used to generate $0 - 10\%$ central Pb-Pb events at $\sqrt{s_{\mathrm{NN}}} = 2.76$ TeV. We found that the extracted identified associated hadron spectra for charged pion, kaon, and proton show identified trigger-hadron dependent splitting. Moreover, the identified trigger-hadron dependent correlation functions vary in different p_T bins, which may show the presence of collective/nuclear effects.

1. Introduction

Charged hadron-hadron correlations have been extensively studied since the ISR and S$p\bar{p}$S experiments in the mid '70's [1]. These measurements have provided a good basis for investigating fragmentation processes. Well before jet reconstruction methods were born, these correlation measurements have been applied more recently in relativistic heavy-ion collisions where full jet reconstruction is challenging, especially at low jet energies. The focus of these studies was to investigate medium effects such as back-to-back jet suppression and modifications of the fragmentation process. Additionally, the discovery of long-range correlation structures in p-p, p-Pb and Pb-Pb collisions at the LHC was based on two-particle correlation measurements [2, 3, 4]. Various hadronization and hydrodynamical models of the space-time evolution of the hot and dense color medium, which have been employed to describe these phenomena, rely heavily on assumed parton and hadron distributions. Nevertheless, the measured hadron spectra and extracted fragmentation functions are integrated distributions which mix up all possible contributions of parton-hadron channels via the convolution of initial-state, QCD and final state effects of high energy collisions [5]. The only way to decouple the specific hadronization contributions is to test the conservation of quantum numbers by tracing the baryon number and flavor-specific emission patterns of hadrons from the partonic medium.

This study focuses on the identified-hadron correlations, which is related to the hadronization mechanisms. Here, the baryon-to-meson ratio (for light and strange particles) is enhanced by more than a factor of three in Pb-Pb collisions compared to p-p collisions [6]. Since this unique kinematic range separates the low momentum thermal bulk production from the parton fragmentation dominated regime at high momenta, the expected production mechanism is of particular interest. On the other hand, the question of quantum number conservation during the phase transition from the partonic to the hadronic world can be addressed only through correlations beyond the simple baryon-to-meson ratios.

In this short review we investigate the strength of the expected effects revealed by PID-triggered hadron-hadron correlations in high-energy p-p and Pb-Pb collisions at LHC energies in the mid-rapidity region ($|\eta| < 1$). This region represents the sensitive geometrical areas of the RHIC/LHC detectors, that are common to STAR, PHENIX, ALICE, ATLAS and CMS.

2. Monte Carlo simulations of identified hadron-hadron azimuthal correlations in p-p and Pb-Pb collisions

2.1. Monte Carlo samples used for the analysis

In this study we focused on the generated identified particles: $\pi^{\pm}$, $K^{\pm}$, p, $\bar{p}$, and the charged hadrons ($h^{\pm}$) as a reference.

Event generators PYTHIA 6.4 [7] and HIJING 1.36 [8] were used for the production of p-p and Pb-Pb event samples. The Perugia-0 tune [9] was chosen for PYTHIA. HIJING event generation included the simulated effects of quenching and shadowing. The generated data include 200 million p-p events at 7 TeV and 4 million central ($0 - 10\%$) Pb-Pb events at 2.76 ATeV. All charged final-state particles were kept for analysis from the Monte Carlo events, nevertheless the trigger particles and associated particles used for the two-particle correlations measurements were limited in pseudorapidity. Trigger particles with $|\eta_{trig}| > 0.5$ and the associated particles with $|\eta_{assoc}| > 1$ were rejected. In order to exclude the low-momentum region from the bulk effects, the transverse momenta of the trigger particles were considered above $p_{T,trig} > 2$ GeV/c in all the cases. The associated particle momenta were chosen such that $p_{T,assoc} < p_{T,trig}$ to avoid double counting. The uncorrelated background was subtracted by the ZYAM method [10].

2.2. The definition of the associated per trigger yield

Measuring quantum number conservation requires constraints both on the trigger and the associated particle directions to tighten the sensitivity of the measurement. The associated per trigger yield is defined with the following quantity as a function of pseudorapidity difference $\Delta\eta = \eta_{trig} - \eta_{assoc}$ and azimuthal angle difference $\Delta\phi = \phi_{trig} - \phi_{assoc}$:

$$\frac{\mathrm{d}^2 N}{\mathrm{d}(\Delta\eta)\mathrm{d}(\Delta\phi)}(\Delta\phi, \Delta\eta, p_{T,trig}, p_{T,assoc}) = \frac{1}{N_{\mathrm{trig}}} \cdot \frac{\mathrm{d}N_{\mathrm{assoc}}}{\mathrm{d}(\Delta\eta)\mathrm{d}(\Delta\phi)}\ , \tag{1}$$

where N_{trig} and N_{assoc} is the number of trigger and associated particles, respectively. This quantity was measured in several $p_{T,trig}$ and $p_{T,assoc}$ intervals. The azimuthal angle correlations in this study are always projected within the pseudorapidity difference $|\Delta\eta| < 1$. The associated per trigger yield on the near-side was extracted from the $\Delta\phi$ interval $|\Delta\phi| < \pi/2$ and the away-side from the $\Delta\phi$ interval $\pi/2 < \Delta\phi < 3\pi/2$.

2.3. The PID-associated spectra

To test the quantum number conservation with $p_{T,assoc}$ the identified associated spectra of the associated particles (hereafter called PID-associated spectra) have been plotted. We have investigated the PID-associated raw spectra using $\pi^{\pm}$, $K^{\pm}$, p, $\bar{p}$, and charged hadron $h^{\pm}$ triggers. Figure 1 shows the raw (unnormalized) identified particle yields $\mathrm{d}N^{PID}_{assoc}/\mathrm{d}p_{T,assoc}$

for the identified associated particles up to $p_{T,assoc} < 25$ GeV/c. The PID-associated spectra have been plotted for positively charged pions, as trigger particles, selected from the transverse momentum range 2 GeV/c $< p_{T,trig} <$ 25 GeV/c. The upper row (Figure 1 *a* and *b*) shows the 7 TeV p-p PYTHIA results and the lower row (Figure 1 *c* and *d*) shows the 0 – 10% central Pb-Pb collisions at 2.76 *A*TeV from HIJING. The left column shows the spectra extracted from the near-side and the right column shows the spectra extracted from the away-side. The PID-associated spectra of the different associated particle species show the expected evolution with the transverse momentum. The yields of the PID-associated spectra significantly decrease with the selection of charged pion, kaon and proton triggers.

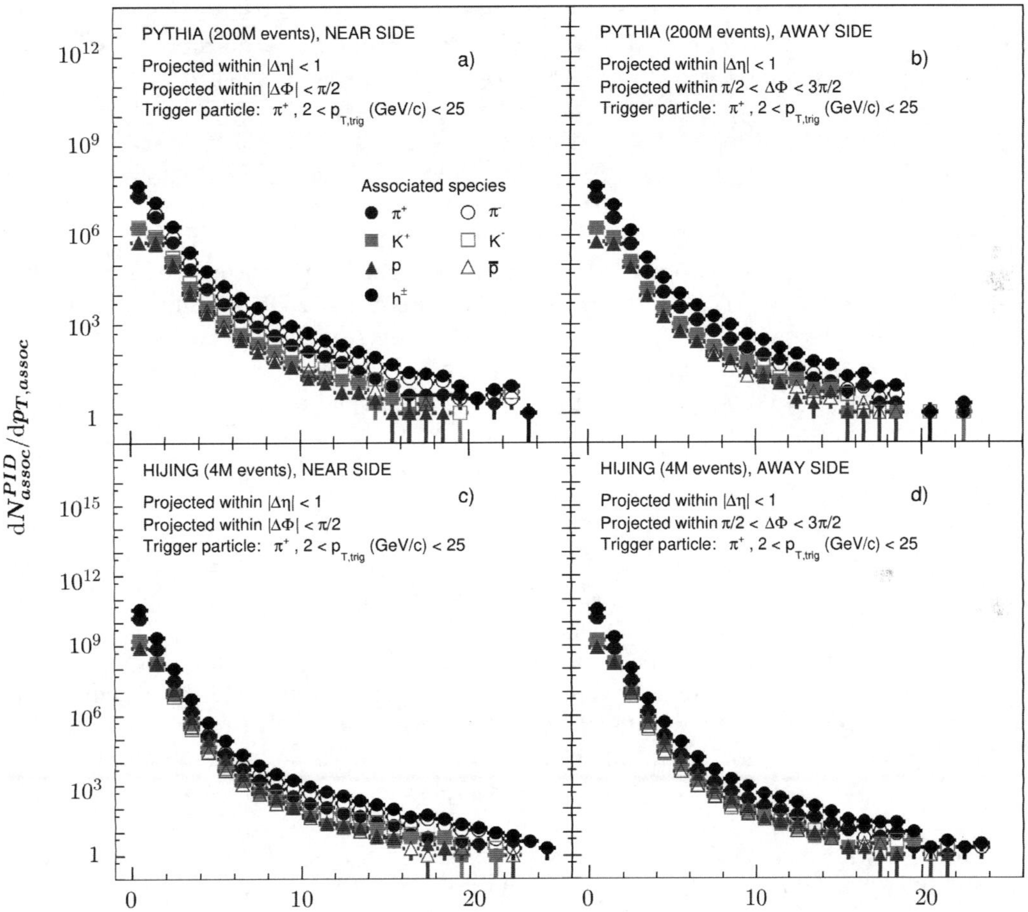

Figure 1. (Color online) Identified associated particle spectra in correlation with positively charged pions (π^+). Open symbols indicate the positive particles and the filled symbols indicate the negative particles. The upper row (*a* and *b*) shows the PYTHIA simulation in p-p and the lower row (*c* and *d*) shows the 0 – 10% central Pb-Pb HIJING events. The left column corresponds to the near side ($\pi/2 < |\Delta\phi|$) correlations and the right column corresponds to the away-side ($\pi/2 < \Delta\phi < 3\pi/2$) correlations.

Journal of Physics: Conference Series **589** (2015) 012001 doi:10.1088/1742-6596/589/1/012001

2.4. Identified particle ratios

At the Monte Carlo event generator level, the basic conservation laws of the quantum numbers – such as charge (C), baryon number (B), and strangeness (S) – are fulfilled and reflected in the PID-associated spectra after the hadronization process. To extract and enhance the expected quantum number conservation effects, we use the following ratio:

$$\frac{\mathrm{d}N_{assoc}^{PID}}{\mathrm{d}p_{T,assoc}} \Big/ \frac{\mathrm{d}N_{assoc}^{hadron}}{\mathrm{d}p_{T,assoc}} \tag{2}$$

which is the PID-triggered to charged hadron-triggered ratio of the PID-associated particle spectra, that is the case when we plot the associated spectra having an identified trigger compared to the one when we have a unidentified (charged hadron) trigger. Furthermore, it can be understood as a specific normalization to the hadron-triggered associated spectra.

In Figure 2 the ratios for six identified trigger / identified associated particle species are shown: π^+/π^+, π^+/π^-, K^+/K^+, K^+/K^-, p/p, and $p/\bar{p}$. The Fig. 2 *a* shows the near-side ratios in p-p events. A visible splitting effect is observed for all the trigger species starting from the first $p_{T,assoc}$ bin up to $10-12$ GeV/c. The largest splitting is observed in the unlike-sign correlation, $p/\bar{p}$, where there is approximately 10-times more anti-protons than protons above a specific $p_{T,assoc}$ bin, that is the splitting strength is about an order of magnitude larger than the like-sign correlation p/p. The splitting, in case of unlike-sign pairs, for K^+/K^- is weaker than in the $p/\bar{p}$ but still comparable with it but the π^+/π^- shows very weak splitting.

In contrast, the away-side p-p ratios, in Fig. 2 *b*, show an equal ratio for the pions and protons (compared to the near-side in Fig. 2 *a*). Conversely, kaons show a weak splitting in a limited transverse momentum range with respect to the near-side. The baryon number and charge is conserved and leads to highly correlated distributions in the same cone, since the fragmentation process was not modified by the media. These qualitative observations show that the strength of the quantum number conservation effects is increasing in the order of charge (C), strangeness (S) and baryon number (B) respectively. In case of hadron correlations with multiple quantum number conservation (C, B, S, ... etc.) this effect will be stronger.

Fig. 2 *c* and *d* exhibit a peculiar pattern for the case of HIJING on the near- and away-side in the $p_{T,assoc} = 2-8$ GeV/c range for charged kaons and protons, where the baryon/meson anomaly was observed at RHIC and at the LHC [11], followed by the reversal of the splitting trend at higher transverse momenta. The charged pions show the splitting pattern only on the near-side.

The lack of the splitting effect below 2 GeV/c may be attributed to the fact that in HIJING the embedded mini-jet production starts to appear around this scale. On the away-side (Fig. 2 *d*), it can be seen that the effect is opposite to the PYTHIA (Fig. 2 *b*), there are more like-sign particles coming from that region. Although it is much smaller than in the near-side, but not negligible. Moreover, we believe that the observed splitting of the protons and anti-protons presumably related to the embedded parametrization of baryon number and charge in HIJING. Further studies are ongoing to characterize better the observed splitting effects that provide an interesting contrast point between various MC event generators and experimental results from the LHC.

2.5. Identified particle ratios in different trigger p_T bins

In the previous subsection the associated particle spectra were studied in the case that the trigger particles have p_T in a wide range (2 GeV/c $< p_{T,trig} <$ 25 GeV/c). In this section we investigate the $p_{T,trig}$ dependence of the spectra. We plot the differences between the like-sign and unlike-sign particle pairs for the normalized yields $(\mathrm{d}N_{assoc}^{PID}/\mathrm{d}p_{T,assoc})/(\mathrm{d}N_{assoc}^{hadron}/\mathrm{d}p_{T,assoc})$

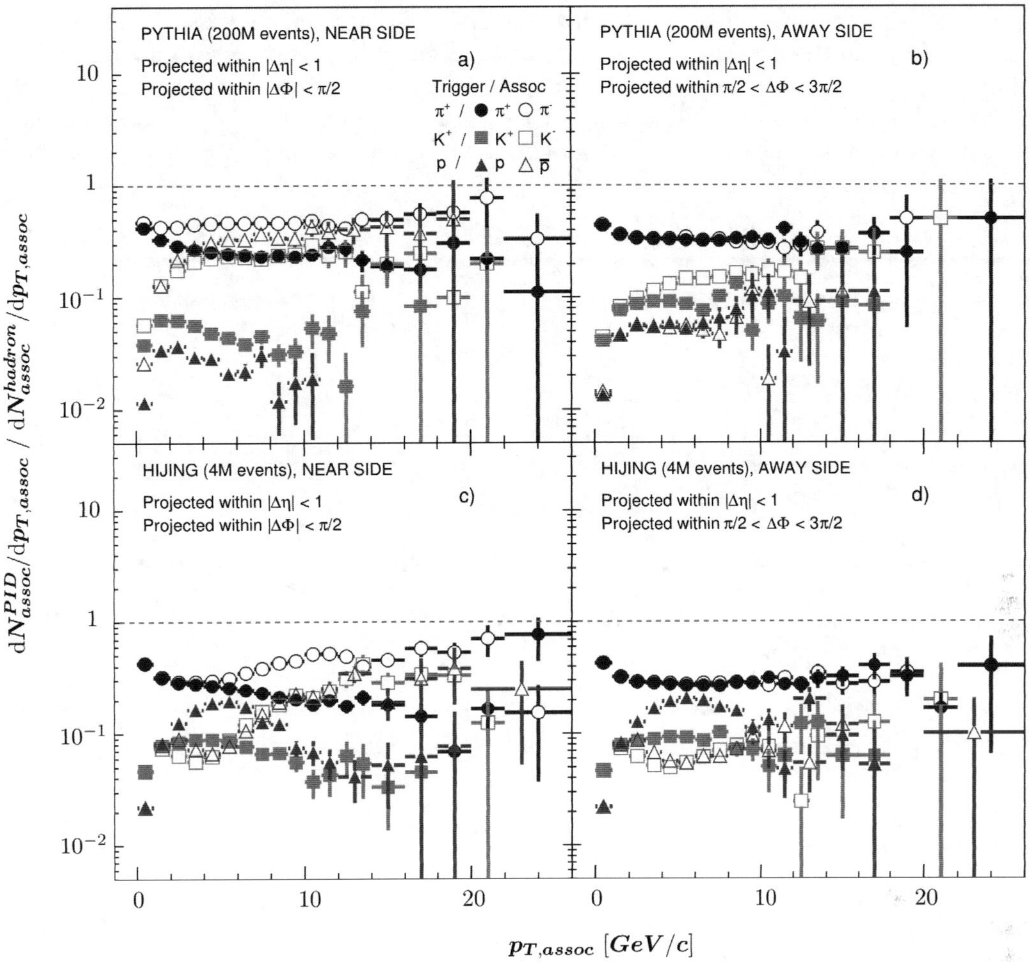

Figure 2. (Color online) PID-triggered associated particle yields relative to the charged hadron-triggered associated yields. Open symbols indicate the positive particles and the filled symbols indicate the negative particles. The three different cases for the trigger/associated particle species are the unlike-sign pairs of π^+/π^-, K^+/K^- and $p/\bar{p}$. The upper row (a and b) shows the PYTHIA simulation in p-p and the lower row (c and d) shows the $0-10\%$ central Pb-Pb HIJING events. The left column corresponds to the near side ($\pi/2 < |\Delta\phi|$) correlations and the right column corresponds to the away-side ($\pi/2 < \Delta\phi < 3\pi/2$) correlations.

as plotted in Fig. 2. Since positively charged triggers were choosen we came to the definition below in order to plot these ratios higher than one:

$$\left(\frac{\mathrm{d}N^{PID}_{assoc}}{\mathrm{d}p_{T,assoc}} \Big/ \frac{\mathrm{d}N^{hadron}_{assoc}}{\mathrm{d}p_{T,assoc}}\right)^{(+/-)} \Big/ \left(\frac{\mathrm{d}N^{PID}_{assoc}}{\mathrm{d}p_{T,assoc}} \Big/ \frac{\mathrm{d}N^{hadron}_{assoc}}{\mathrm{d}p_{T,assoc}}\right)^{(+/+)} = \left(\frac{\mathrm{d}N^{PID}_{assoc}}{\mathrm{d}p_{T,assoc}}\right)^{(+/-)} \Big/ \left(\frac{\mathrm{d}N^{PID}_{assoc}}{\mathrm{d}p_{T,assoc}}\right)^{(+/+)} \tag{3}$$

where $(+/-)$ and $(+/+)$ denote the unlike-sign and like-sign trigger/associated particle pairs, respectively. This ratio gives a more detailed insight into the evolution of the particle yield differences between the positive and negative associated particles as a function of $p_{T,assoc}$. The ratios defined by Eq. (3) are plotted for four different $p_{T,trig}$ trigger bins on the horizontal axis of Fig. 3. In Fig. 3 *a* it can be observed how the splits – in agreement with Fig. 2 – evolve as a function of $p_{T,assoc}$ in the near-side cone in PYTHIA events. In general, the splitting effect is the largest and growing for the case when we look at the differences between the p and $\bar{p}$. Concerning strange particles it seems to saturate at higher $p_{T,assoc}$. For pions the relative difference is about factor of $2 - 3$ at most. There is almost no change in the ordering of the species in the other $p_{T,trig}$ bins, $3 - 7$ GeV/c, $7 - 10$ GeV/c, and $10 - 20$ GeV/c, and the shapes remain at the same level.

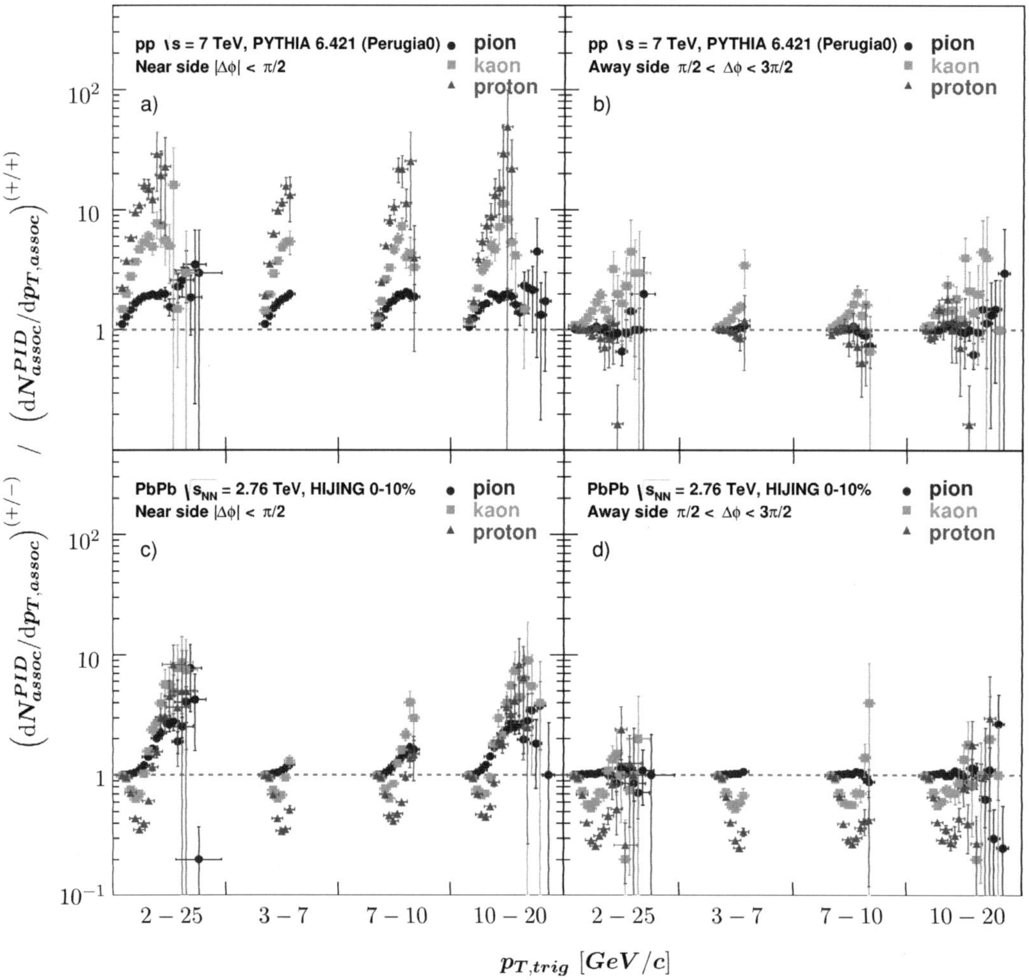

Figure 3. (Color online) Yield differences between the oppositely charged trigger/associated particle pairs. The yield enhancement is relative to the positively charged trigger particles. The upper row (*a* and *b*) shows the PYTHIA simulation in p-p and the lower row (*c* and *d*) shows the $0 - 10\%$ central Pb-Pb HIJING events. The left and right panels correspond to the near- and away-side cases, respectively.

Concerning the away-side, in Fig. 3 *b*, pions and protons with opposite charges are generated with the same yield but there can be a marginal difference for the kaons as it was shown in Fig. 2 *b*. For HIJING-generated Pb-Pb events in the $0 - 10\%$ central region, Fig. 3 *c* and *d*, the kaons and the protons show an reverse trend with respect to the PYTHIA events. In the near-side case, the spectrum is different for the first $p_{T,assoc}$ bins in all cases for the trigger

$p_{T,trig}$ bins and after that rises above unity. The number of K's produced (compared to π's) is proportional to $p_{T,assoc}$.

This trend is unexpected and completely opposite to the PYTHIA on the away-side for all the $p_{T,trig}$ bins. Moreover, the effect manifests with similar magnitude within any $p_{T,trig}$ ranges. In summary, these effects are observed in the central Pb-Pb collisions within the $3-10$ GeV/c transverse momentum range, where the Cronin effect [12] and the baryon/meson anomaly act. Conversly these effects are not seen in the bulk, as indicated by the plot.

2.6. Testing conservation laws by $\Delta\phi$ correlations

In Fig. 4 the $\Delta\phi$-distributions given by Eq. (1) of identified two-particle correlations are shown for unlike-sign trigger/associated particle pairs in different $p_{T,trig}$, $p_{T,assoc}$ bins extracted from PYTHIA events. Recall that the spectra, mentioned previously, are baseline-subtracted by applying the ZYAM-method. The columns refer to the different species of the trigger/associated particle pairs, going from left to right we plotted π^+/π^-, K^+/K^- and $p/\bar{p}$. The rows correspond to the different settings of the $p_{T,trig}$ bins, going from top to bottom they are 3 GeV/c $< p_{T,trig} <$ 7 GeV/c, 7 GeV/c $< p_{T,trig} <$ 10 GeV/c, and 10 GeV/c $< p_{T,trig} <$ 20 GeV/c.

The near- and away-side peaks show the expected evolution of $p_{T,assoc}$. Higher trigger momentum $p_{T,trig}$ yields a narrower near-side peak, whereas higher associated momentum particles $p_{T,assoc}$ result in a lower associated particle yield. Nevertheless, the amplitude of the correlation peak stays approximately constant on the away-side as a function of $p_{T,trig}$. If we compare the columns, that is the different trigger/associated particle pairs, we can observe that on the near-side, the ratio of the number of the associated particles to the trigger particles is less for strange particles with respect to the baryons compared to the pions. There is a hint that the same effect is visible on the away-side, but a more thorough study is needed.

In Fig. 5 the distributions for the Pb-Pb events, generated by HIJING, show the same trend that is seen on the near-side for the PYTHIA events.

The difference between PYTHIA and HIJING in terms of the chosen trigger/associated species is evident, practically there is only a small flavour and charge correlation effect in HIJING at all studied $p_{T,trig}$ momentum bin. Moreover, the number of associated particles for kaons and protons has the same order of magnitude on the near-side as well as on the away-side.

3. Summary and Outlook

Monte Carlo studies of two-particle azimuthal correlations have been performed with identified trigger particles and identified associated particles in p-p collisions at $\sqrt{s} = 7$ TeV and in Pb-Pb collisions at $\sqrt{s_{\rm NN}} = 2.76$ ATeV at mid-rapidity. As for the first verification of the generated samples the identified hadrons were extracted from the raw associated particle spectra: π^+, π^-, K^+, K^-, p and $\bar{p}$ with fixed positively charged pion (π^+) trigger in two separated $\Delta\phi$ regions, near- and away-side. Further investigation was needed to look deeper into the validity of the quantum number conservations of correlated particle pairs. For this reason we plotted the identified associated spectra for like-sign and unlike-sign trigger/associated particle pairs compared to the charged hadron identified associated spectra. We obtained conclusive results for PYTHIA: the PID-triggered-to-hadron-triggered associated spectra on the near-side show a significant splitting between the particles and anti-particles which have the largest effect for the baryon triggers, but is absent on the away-side. Qualitatively the difference between the baryon triggers and the pions is about one order of magnitude. For the HIJING simulation the results show an opposite trend to the PYTHIA up to $p_{T,assoc} = 7-10$ GeV/c and these effects are not vanishing on the away-side except for the pions. The splitting in Pb-Pb collisions starts at slightly larger $p_{T,assoc}$ threshold than in p-p because of the embedded physics in HIJING (for example the minijet production), but further investigation is needed with real data to confirm

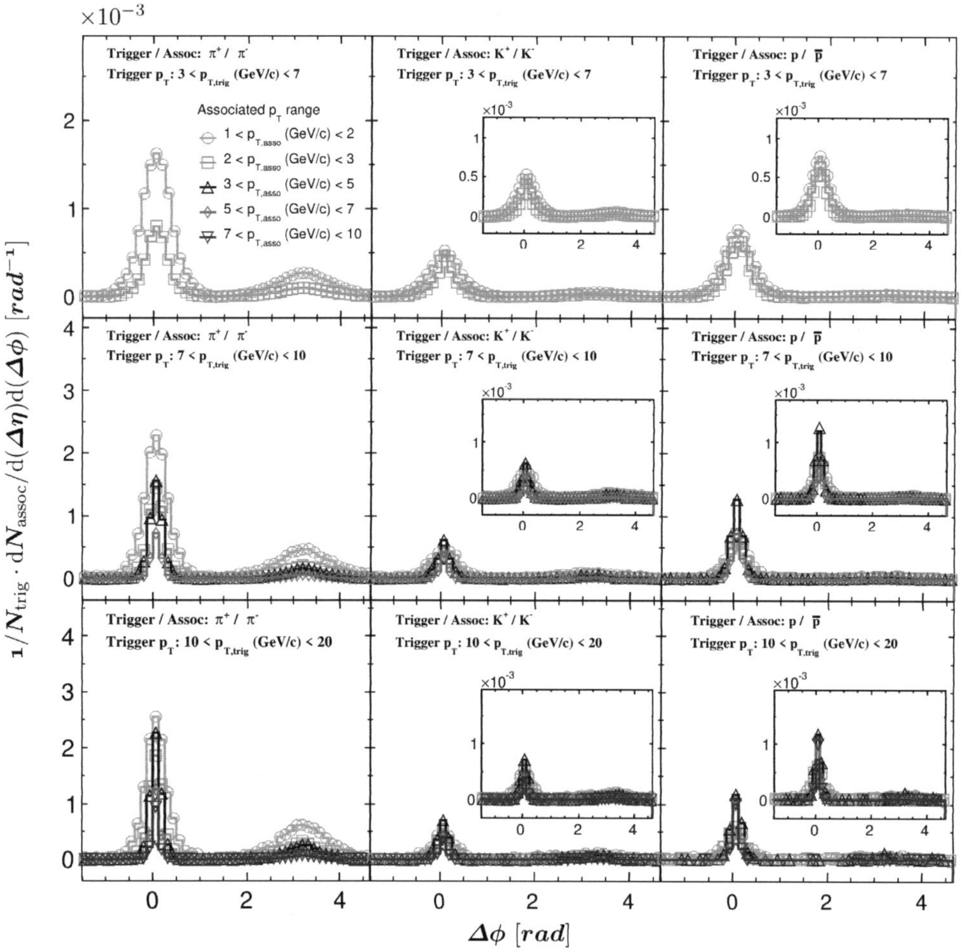

Figure 4. (Color online) PYTHIA simulation of the p-p collisions at $\sqrt{s} = 7$ TeV. The associated per trigger yields are plotted for unlike-sign trigger/associated particle pairs: π^+/π^-, K^+/K^- and $p/\bar{p}$ in different $p_{T,trig} - p_{T,assoc}$ bins. The extracted distributions are integrated in the pseudorapidity difference range $|\Delta\eta| < 1$. The different columns and rows correspond to different species of trigger/associated particle pairs and trigger momentum bins $p_{T,trig}$, respectively. The insets correspond to magnified portion of the y-axis of the plotted distribution for better visibility.

this observation. To conclude: collective/nuclear effects are visible in the studied $3 - 10$ GeV/c momentum range.
There is no significant change between the yields of the unlike-sign particle pairs as we change the trigger $p_{T,trig}$ bins. Finally, we showed the PID-triggered azimuthal $\Delta\phi$ correlation spectra for the different trigger species. We have seen more associated protons than kaons on the near-side of the p-p collisions which does not depend on the applied trigger momentum $p_{T,trig}$ cuts. In Pb-Pb collisions the probability to generate protons or kaons as associated particles on the near-side is about equal.

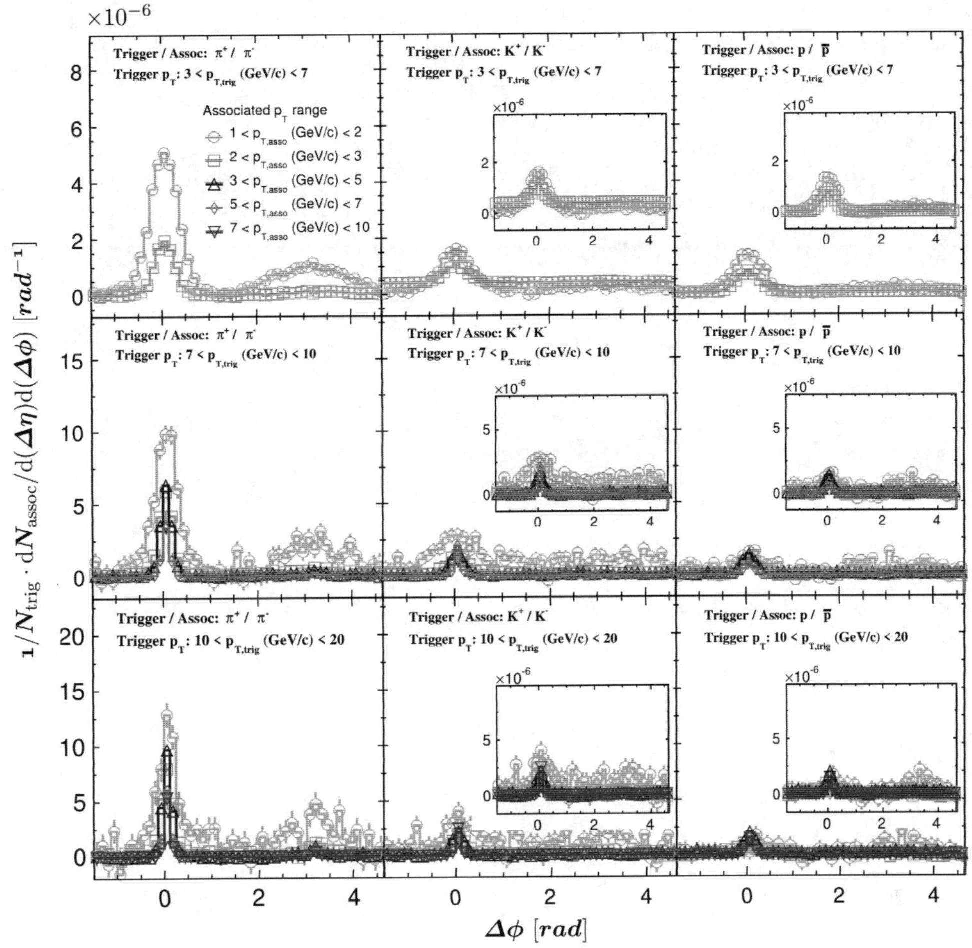

Figure 5. (Color online) HIJING simulation of Pb-Pb collisions at $\sqrt{s_{\rm NN}} = 2.76$ TeV. The associated per trigger yields are plotted for unlike-sign trigger/associated particle pairs: π^+/π^-, K^+/K^- and $p/\bar{p}$ in different $p_{T,trig}$–$p_{T,assoc}$ bins. The extracted distributions are integrated in the pseudorapidity difference range $|\Delta\eta| < 1$. The different columns and rows correspond to different species of trigger/associated particle pairs and $p_{T,trig}$ trigger momentum bins, respectively. The insets correspond to magnified portion of the y-axis of the plotted distribution for better visibility.

Acknowledgement

This work was supported by Hungarian OTKA grants NK778816, NK106119, K104260, and TET 12 CN-1-2012-0016. Author Gergely Gábor Barnaföldi also thanks the János Bolyai Research Scholarship of the Hungarian Academy of Sciences.

References

[1] M. G. Albrow *et al.*, Phys. Lett. B **65** (1976) 295.
[2] S. Chatrchyan *et al.* [CMS Collaboration], [arXiv:1310.8651 [nucl-ex]].
[3] B. Abelev *et al.* [ALICE Collaboration], Phys. Lett. B **719** (2013) 29 [arXiv:1212.2001 [nucl-ex]]; B. B. Abelev *et al.* [ALICE Collaboration], Phys. Lett. B **726** (2013) 164 [arXiv:1307.3237 [nucl-ex]].
[4] G. Aad *et al.* [ATLAS Collaboration], JHEP **1311** (2013) 183 [arXiv:1305.2942 [hep-ex]].

[5] R. K. Ellis *et al.* QCD and Collider Physics, Cambridge Monographs on Particle Physics, Nuclear Physics and Cosmology, 2003, p. 452 (ISBN: 9780521545891)
[6] B. B. Abelev *et al.* [ALICE Collaboration], Phys. Rev. Lett. **111** (2013) 222301 [arXiv:1307.5530 [nucl-ex]].
[7] T. Sjostrand *et al.* PYTHIA 6.4 Physics and Manual, JHEP 05 (2006) 026. arXiv:hep-ph/0603175
[8] X.-N. Wang, M. Gyulassy, Phys. Rev. **D44** (1991), 3501 – 3516.
[9] P. Z. Skands, Phys. Rev. **D82** (2010) 074018 arXiv:hep-ph/1005.3457
[10] A. Adare *et al.* [PHENIX Collaboration], Phys. Rev. C **78** (2008) 014901 [arXiv:0801.4545 [nucl-ex]];
[11] H. Appelshauser, J. Phys. G **38**, 124014 (2011)
[12] Antreasyan, D. *et al.* Phys. Rev. Lett. **38** (1977)
[13] *A Very High Momentum Particle Identification Detector (VHMPID) for ALICE, Letter of Intent*, arXiv:1309.5880 [nucl-ex]

Journal of Physics: Conference Series **589** (2015) 012002 doi:10.1088/1742-6596/589/1/012002

Overview of STAR results on correlations, jets and heavy-flavor production

Jana Bielčíková for the STAR Collaboration

Nuclear Physics Institute ASCR, Na Truhlářce 39/64, 180 86 Praha, Czech Republic

E-mail: jana.bielcikova@ujf.cas.cz

Abstract. Measurements of jets and heavy-flavor production play an important role in understanding properties of hot and dense nuclear matter created in high energy heavy-ion collisions. As direct measurements of jets are difficult due to large underlying background, jet properties can be also studied via di-hadron and multihadron correlations. In this contribution to these proceedings, an overview of recent STAR results on correlations, jet-hadron correlations, and heavy-flavor production in Au+Au collisions at top RHIC energy ($\sqrt{s_{NN}}$ = 200 GeV) is given. In order to draw quantitative conclusions on properties of the hot and dense nuclear matter created in Au+Au collisions, reference measurements in elementary proton-proton collisions and d+Au collisions are essential to estimate contributions due to cold nuclear matter effects and are discussed as well.

1. Introduction

Exploration of nuclear matter under extreme conditions of temperature and energy density is on the forefront of experimental and theoretical research pursued in nuclear physics. By combining multiple experimental signatures, experiments at the Relativistic Heavy Ion Collider (RHIC) at Brookhaven National Laboratory have concluded that a new state of nuclear matter was created in central Au+Au collisions at the center of mass energy $\sqrt{s_{NN}}$ = 200 GeV per nucleon-nucleon pair [1–4] with properties of a strongly coupled liquid with a very low ratio of shear viscosity to entropy. The nuclear matter created at RHIC is opaque to the passage of energetic quarks and gluons which is reflected in a large suppression of particle production at large transverse momenta (p_T) relative to elementary p+p collisions. This effect, commonly referred as to "jet quenching", is observed for light as well as heavy quark flavors. In this proceedings, an overview is given of recent STAR results on measurements of correlations at intermediate and high p_T, jets and heavy flavor production, the so called hard probes, in p+p, d+Au and Au+Au collisions at top RHIC energy. Latest STAR measurements of identified two-particle correlations as well as direct photon elliptic flow measurements are reported in [5, 6] as presented at this workshop.

2. Di-hadron correlations in d+Au collisions

Long-range correlations in pseudo-rapidity (the so called "ridge") at near side of trigger hadron were discovered at RHIC almost a decade ago [7, 8] and were systematically studied in [9, 10]. The origin of this phenomenon in heavy-ion collisions is most probably connected with initial state fluctuations resulting in triangular flow (v_3) [11]. Recent measurements at LHC energies of di-hadron correlations in small systems, such as p+p [12, 13] and p+Pb [14–16] collisions, revealed

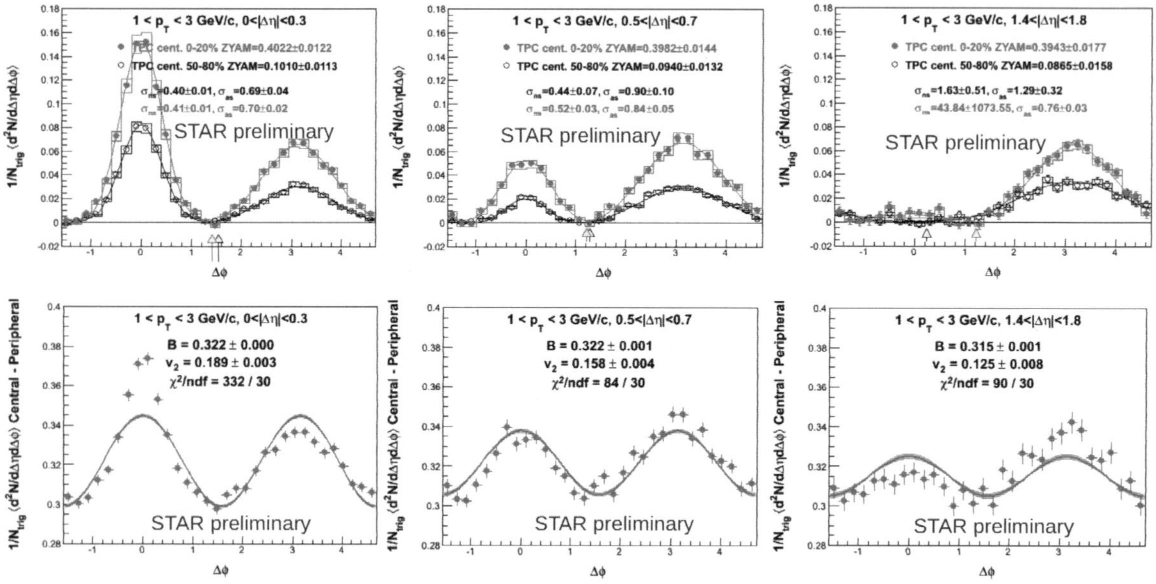

Figure 1. Top row: Di-hadron correlations in three $|\Delta\eta|$ intervals: $0 < |\Delta\eta| < 0.3$ (left), $0.5 < |\Delta\eta| < 0.7$ (middle), $1.4 < |\Delta\eta| < 1.8$ (right) in 50-80% (open black symbols) and 0-20% (close red symbols) d+Au collisions at $\sqrt{s_{NN}} = 200$ GeV. The centrality was determined from charged particle multiplicity in TPC. Both trigger and associated charged particles have $p_T = 1$-3 GeV/c and $|\eta| < 1$. The boxes indicate systematic uncertainties due to ZYAM background subtraction. Bottom row: Correlation patterns left after subtracting central and peripheral correlations functions from the top row. The curves correspond to $B(1 + 2v_2^2\cos(2\Delta\phi))$ fit of the data.

unexpected ridge-like structures as well. In high-multiplicity p+p collisions, an extended ridge-like structure was observed at near side of the trigger hadron at low and intermediate transverse momenta. In p+Pb collisions, when subtracting correlations in low-multiplicity events from high-multiplicty events, two ridge-like structures were observed - one on the near side and one on the away side of the trigger hadron. The yields at both near and away side of the trigger hadron were found to be within uncertainties equal in the studied kinematic range and the widths show no significant evolution with collision multiplicity compared to p+p collisions. The "ridge-like" pattern is much stronger in p+Pb collisions keeping the same charged particle multiplicity as in p+p collisions.

This novel ridge structure triggered intense theoretical discussions and its origin is not fully understood. Various mechanisms proposed to explain the origin of the ridge in high-multiplicity p+p and p+Pb collisions include color field connections in the longitudinal direction [17–19] jet-medium [20] and multi-parton induced interactions [21, 22], and collective effects in high multiplicity p+p and p+Pb collisions [23–26]. On the experimental side the observation of the ridge in cold nuclear matter at LHC lead to a renewed interest in d+Au collisions at RHIC in an effort to search for similar effects at lower collision energy. PHENIX has published their results in [27]. Here we discus recent STAR measurements of di-hadron correlation in d+Au collisions.

The presented analysis is based on d+Au data set measured in 2003 at $\sqrt{s_{NN}} = 200$ GeV. The trigger particles were reconstructed in the Time Projection Chamber (TPC) within pseudorapidity $|\eta| < 1.0$. These trigger particles were then correlated with associated particles

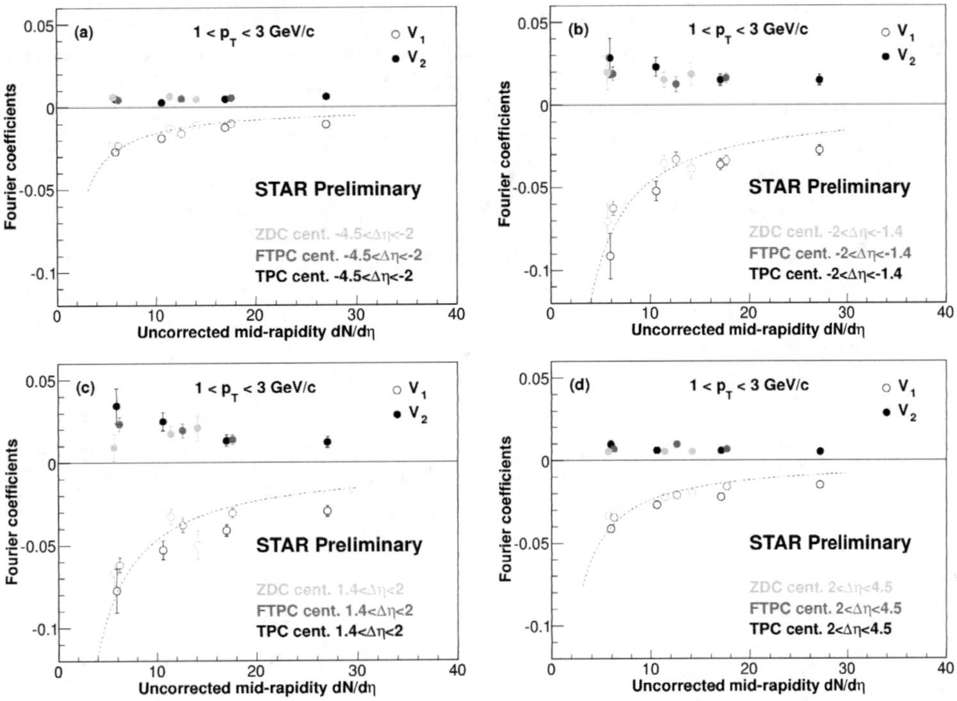

Figure 2. Centrality dependence of Fourier coefficients V_1 and V_2 of di-hadron correlation functions in d+Au collisions at $\sqrt{s_{NN}}$ = 200 GeV for four $\Delta\eta$ ranges (see legend). Results from all three centrality definitions are shown (see legend). Both trigger and associated charged particles have p_T = 1–3 GeV/c and $|\eta| < 1$. The line represents a fit of V_1 by a function proportional to $1/(\mathrm{d}N/\mathrm{d}\eta)$.

from the TPC ("TPC-TPC" correlations) or with those measured in Forward Time Projection Chambers ("TPC-FTPC" correlations) placed at -3.8 < η < -1.8 and 1.8 < η < 3.8, respectively. The correlation functions were normalized per trigger particle and corrected for reconstruction efficiency of associated particles. The two-particle acceptance in $\Delta\eta - \Delta\phi$ was corrected by mixed event technique. The correlations were analyzed as a function of centrality and $\Delta\eta$. To define centrality of d+Au collisions, three methods were used: multiplicity of charged particles in the TPC within $|\Delta\eta| < 0.5$, multiplicity of charged particles within -3.8< η <-1.8 in the FTPC in the Au-beam direction, and neutral energy measured in the Zero Degree Calorimeter (ZDC) also placed in the Au-beam direction.

An example of the corrected di-hadron TPC-TPC correlations for transverse momentum p_T = 1-3 GeV/c in two multiplicity classes (0-20% and 50-80%) and three $|\Delta\eta|$ windows is presented in Figure 1. The upper row shows correlations in the two centrality bins after background subtraction based on the ZYAM method [28]. The systematic errors due to ZYAM subtraction are shown as boxes around the measured data points. In the bottom row of Figure 1 the correlations patterns are displayed after subtracting correlations in the mentioned centrality bins (i.e. central-peripheral). In the latter case, no background subtraction method is needed. Projections of the remaining "central-peripheral" correlation patterns to $\Delta\eta$ separately for near and away-side region in $\Delta\phi$, show a Gaussian-like excess at near side and an approximately uniform away side. For further details as well as HIJING model comparisons we refer reader to results presented at [29].

Journal of Physics: Conference Series **589** (2015) 012002 doi:10.1088/1742-6596/589/1/012002

Next, Fourier analysis of the di-hadron correlations was performed for both TPC-TPC as well as TPC-FTPC correlations. The centrality dependence of the extracted Fourier coefficients $V_n = \langle\cos(n\Delta\phi)\rangle$, where n=1,2 is shown in Figure 2. The V_1 and V_2 coefficients were calculated for all three centrality determination methods. The V_1 coefficient is found to be inversely proportional to charged particle multiplicity $dN/d\eta$ at mid-rapidity. The V_2 coefficient is small and approximately independent of $dN/d\eta$. The physics mechanism which would explain non-zero values of the Fourier coefficients for "central-peripheral" correlations is currently under investigation [29].

3. 2+1 azimuthal correlations

In [30], STAR introduced a novel method for investigating jet-medium interactions via a multi-hadron correlation technique (called "2+1"), where a pair of back-to-back hadron triggers with large transverse momentum is used as a proxy for a di-jet. In this first analysis, the results were reported for nearly-symmetric trigger pairs with the highest momentum threshold of the trigger hadron of 5 GeV/c. The distributions of associated hadrons were investigated in terms of shapes, approximated by a Gaussian peak, as well as per-trigger yields on each trigger side. The properties of the di-jet like correlations for the symmetric trigger case in central Au+Au collisions at $\sqrt{s_{NN}}$ = 200 GeV were found to be similar in both shape and magnitude to those in d+Au collisions. This observation was suggestive of a strong surface (tangential) bias of selected di-jets.

Here we report an extension of this analysis to higher primary trigger thresholds and asymmetric back-to-back triggers as a tool for changing the amount of surface bias [31]. As primary (T1) triggers, clusters measured in the STAR Barrel ElectroMagnetic Calorimeter (BEMC) with the transverse energy of at least 8 GeV were selected. A highest-p_T charged particle, measured in the STAR TPC, in the back-to-back azimuthal region relative to T1 ($|\phi^{\mathrm{T1}} - \phi^{\mathrm{T2}} - \pi| < 0.2$) with $4 < p_T^{\mathrm{T2}} < 10$ GeV/c was then selected as secondary trigger (T2). Following the notation introduced in [30], the $\eta - \phi$ region close to T1 is defined as near-side region and $\eta - \phi$ region close to T2 as the away-side region, respectively. The new analysis takes advantage of the triggering capabilities of the BEMC in Run7. This allowed for significant improvement of the kinematic reach and asymetry. The correlations relative to T1 and T2 triggers were constructed using charged hadrons with transverse momentum p_T^{assoc} =1-10 GeV/c. To control the degree of the surface bias the relative energy balance between T1 and T2 was varied. For the most asymmetric trigger pair selection the T1 trigger had more than twice the energy of its T2 trigger partner.

The energy imbalance between the near- and away-side correlations is quantified by

$$\Delta(\Sigma E_T) = (E_T^{T1} + \Sigma p_T^{assoc,near}) - (p_T^{T2} + \Sigma p_T^{assoc,away}) \quad (1)$$

In Figure 3, the relative energy imbalance between central Au+Au and d+Au data (i.e. $\Delta(\Sigma E_T)^{\mathrm{Au+Au}} - \Delta(\Sigma E_T)^{\mathrm{d+Au}}$), is shown for both the symmetric and asymmetric trigger cases. The errors shown include both statistical and systematic uncertainties. The observed relative energy imbalance for the asymmetric trigger pairs and studied p_T range of associated charged particles is consistent with non-zero values. The relative imbalance reaches the value of ~1.5 GeV/c for the highest p_T associated bin and decreases with lowering p_T of the associated particles. This suggests that the energy missing from the away-side peak at higher p_T^{assoc} is converted into softer hadrons. This observation is in line with expectation that larger asymmetries are related to longer paths traversed in medium and consequently higher energy loss. The measured values of relative energy imbalance were found to be significantly smaller than expected in the path-dependent energy loss model [32], which predicts 3 GeV for the asymmetric trigger case. We note that the imbalance observable is, however, not sensitive to

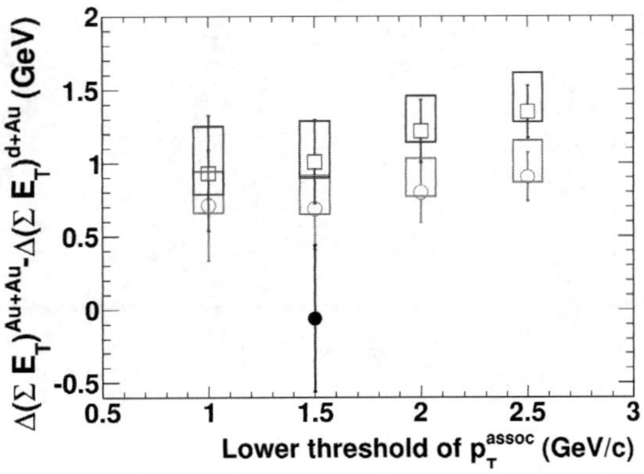

Figure 3. The relative di-jet energy imbalance estimate $\Delta(\Sigma E_T)^{\mathrm{Au+Au}} - \Delta(\Sigma E_T)^{\mathrm{d+Au}}$. The open circles show results for primary triggers in (8-10) GeV, the open squares in (10,15) GeV. The solid data point is the result for symmetric trigger case from [30]. The figure is from [31].

cases when a part of energy deposited to the medium is converted to particle production within the angular selection of the near- and away-side peaks.

4. Jet-hadron correlations

Jet-hadron correlations enable a more direct access to the parton energy in comparison with di-hadron correlations, which use a high p_T trigger hadron as a proxy of jet axis. At the same time, jet-hadron correlations increase the kinematic reach of the correlation measurements. The STAR experiment presented recently a novel study of jet-hadron correlations in p+p and central Au+Au collisions. To enhance contribution of jets, events were selected based on an online high-tower (HT) trigger in the BEMC with an energy deposition in at least one BEMC tower above 5.4 GeV. In addition, a common offline software HT threshold of 6 GeV was required as well. Jets were reconstructed using the anti-k_T recombination algorithm [33] with a resolution parameter $R = 0.4$ from charged tracks detected in the TPC and the transverse energy measured in the BEMC. To suppress background fluctuations, only tracks with $p_T > 2$ GeV/c and BEMC towers with $E_T > 2$ GeV were used to reconstruct jets. Furthermore, the reconstructed jet had to include as one of its constituents a BEMC tower that fired the HT trigger. To avoid double counting, a 100% hadronic correction scheme was applied to account for charged hadron energy deposition in the BEMC.

The jet-hadron correlations reported below [34] were defined as distributions in $\Delta\phi = \phi_{jet} - \phi_{assoc}$, where ϕ_{jet} is the azimuthal angle of the axis of a reconstructed (trigger) jet and the associated particles are all charged hadrons in a given event. The distributions were corrected for single particle tracking efficiency and for detector pair acceptance by event mixing in $\Delta\phi$. To quantify the associated particle yields (Y) and widths (σ) of the near and away-side peaks, the correlation functions were fit with two Gaussians together with a background term modulated by elliptic ($v_2^{assoc}v_2^{jet}$) and triangular ($v_3^{assoc}v_3^{jet}$) Fourier components.

The medium modification of near and away-side peaks was then quantified by studying their Gaussian widths as well as by D_{AA} defined as:

$$D_{AA}(p_T^{assoc}) \equiv Y_{AuAu}(p_T^{assoc}) \cdot \langle p_T^{assoc}\rangle_{AuAu} - Y_{pp}(p_T^{assoc}) \cdot \langle p_T^{assoc}\rangle_{pp}. \quad (2)$$

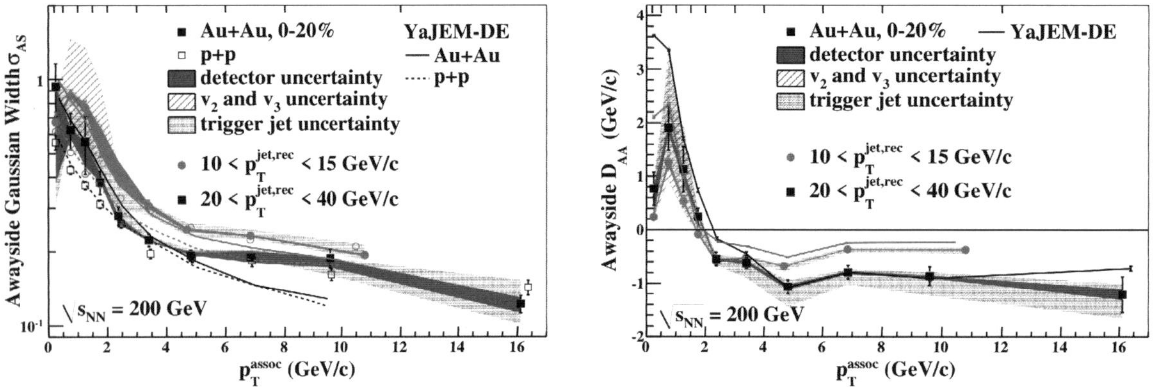

Figure 4. (Left): The Gaussian widths of the away-side jet peak (σ_{AS}) in Au+Au (solid symbols) and p+p (open symbols) for two ranges of $p_T^{jet,rec}$: 10-15 GeV/c (red circles) and 20-40 GeV/c (black squares). (Right) The away-side momentum difference D_{AA} for two ranges of $p_T^{jet,rec}$: 10-15 GeV/c (red circles) and 20-40 GeV/c (black squares). The data are compared to YaJEM-DE model calculations (solid and dashed lines) from [35].

and ΣD_{AA} which is a measure of the energy balance over the entire p_T^{assoc} range:

$$\Sigma D_{AA} \equiv \sum_{p_T^{assoc} bins} D_{AA}(p_T^{assoc}) \tag{3}$$

Any deviations from zero would indicate in-medium modification of fragmentation patterns.

As the trigger jet is expected to have a surface bias, the recoiling parton will travel a significant distance through the medium. This should be reflected in an enhanced partonic energy loss and quenching on the away side. The away-side Gaussian widths (σ_{AS}) and D_{AA} are shown in Figure 4 for two ranges in reconstructed jet transverse momentum, $p_T^{jet,rec}$. The away-side widths at high p_T^{assoc} are on average the same in both p+p and Au+Au collisions pointing to the fact that high p_T fragments are not deflected in the medium. The widths at low p_T^{assoc} are suggestive of in-medium broadening, but the systematic errors due to flow subtraction are large and measurements of jet flow are necessary before any quantitative conclusions can be drawn. The away-side D_{AA} exhibits suppression of high p_T^{assoc} hadrons and enhancement of low-p_T^{assoc} fragments in Au+Au. This indicates softening of fragmentation in Au+Au relative to p+p collisions. A detailed study of ΣD_{AA} values [34] also indicated that the high p_T^{assoc} suppression is largely balanced by the low-p_T^{assoc} enhancement.

The data in Figure 4 are compared to theoretical calculations from YaJEM-DE [36] Monte Carlo model of in-medium shower evolution. The model incorporates both radiative and collisional energy loss and is successfull in describing a range of high p_T observables at RHIC. The redistribution of energy from high to low-p_T charged particles which remain correlated with the jet is quantitatively consistent with the YaJEM-DE model expectations.

5. Open charm production

Particles containing a charm (c) or a beauty (b) quark, the so called heavy flavors, provide due to their large mass better sensitivity to medium properties than light quarks. However, the rate of heavy flavor production is much smaller therefore requiring large data samples. In recent years, the increased luminosity of RHIC and data taking rates allowed to significantly improve statistics and thus investigate in a systematic way the charm production. The charm production

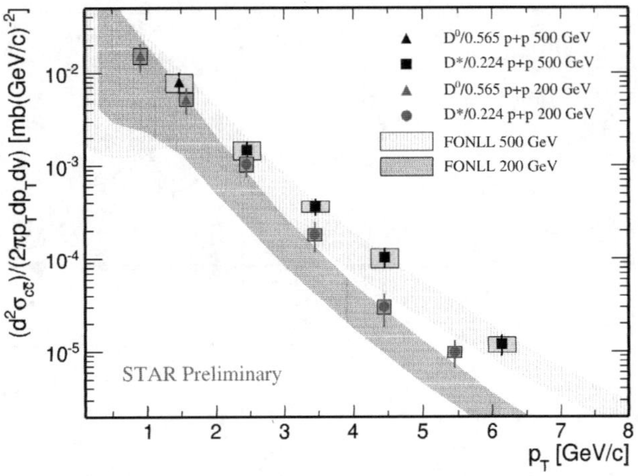

Figure 5. Transverse momentum dependence of $c\bar{c}$ production cross section calculated from D^0 and D^* yields in p+p collisions at $\sqrt{s}$ = 200 GeV [37] and 500 GeV [38]. The data are compared to FONLL calculations [39].

can be indirectly measured via electrons from heavy-flavor decays, also commonly referred to as "non-photonic electrons" (NPE) produced from semileptonic heavy-flavor decays. A direct access to charm production requires reconstruction of e.g. D mesons containing the c quark.

For measurements in heavy-ion collisions it is essential to understand first heavy flavor production in elementary collision systems such as p+p. Here we feature the most recent STAR measurement of $c\bar{c}$ production cross section from hadronic decays of D^0 and D^* mesons in p+p collisions at $\sqrt{s}$ = 200 GeV [37] and 500 GeV [38]. The corresponding charm spectra are shown in Figure 5 and are compared to the Fixed-Order-Next-to-Leading-Logarithm (FONLL) calculations from [39]. The FONLL calculations are found to agree well with data given current experimental and theoretical uncertainties. These measurements constitute an important baseline for similar measurements in heavy-ion collisions.

It was originally expected that in the medium, due to a suppression of gluon radiation at small angles known as the "dead cone" effect [40], heavy quarks will exhibit less suppression than light quarks. However, already the first measurements in central Au+Au collisions at RHIC revealed that production of NPE shows surprisingly similar level of suppression as that of light quarks [41–43]. Several mechanisms have therefore been suggested to explain the large NPE suppression in addition to radiative energy loss. Models, which also incorporate collisional energy loss and energy loss due to in-medium dissociation of heavy flavor D and B mesons, were proven to be more successful in the description of NPE R_{AA}, see e.g. [44–46].

Here we demonstrate the large suppression of charm production on the newest STAR data [38, 47]. The left panel of Figure 6 shows the nuclear modification factor of NPE and D^0 mesons in 0-10% most central Au+Au collisions at $\sqrt{s_{NN}}$ = 200 GeV. As it can be seen, the suppression of NPE and charm mesons is consistent with each other and with that of pions. In the right panel of Figure 6 a closer investigation of R_{AA} of D^0 mesons in central as well as minimum bias Au+Au collisions in the transverse momentum range $0 < p_T < 6$ GeV/c is shown. The R_{AA} has a peaked structure with a maximum between p_T = 1-3 GeV/c. At higher momenta ($p_T > 3$ GeV/c), there is a strong suppression of D^0 yields which increases with centrality. This indicates a large energy loss of charm quarks traversing through the medium. The data are also compared with two transport model calculations [48, 49] based on the relativistic Langevin

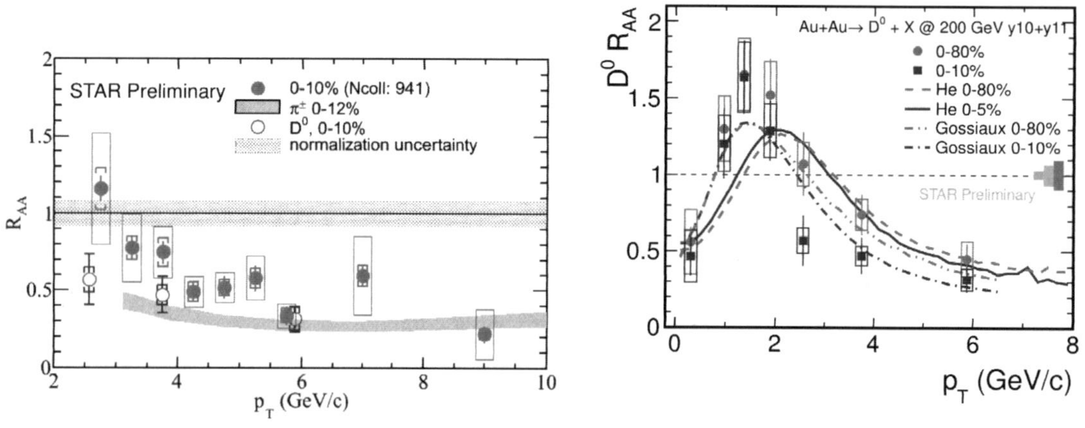

Figure 6. (Left) Transverse momentum dependence of nuclear modification factor of non-photonic electrons (full circles) and D^0 mesons (open circles) in 0-10% most central Au+Au collisions at $\sqrt{s_{NN}}$ =200 GeV. For comparison the nuclear modification factor of pions is shown as well (grey band). (Right) Nuclear modification factor of D^0 meson vs p_T for 0-10% and 0-80% centralities in Au+Au collisions at $\sqrt{s_{NN}}$ = 200 GeV. The data are compared to two transport model calculations (see legend). Error bars on the data points are statistical while boxes represent systematic uncertainties. The figures are from [38].

simulation. The calculations describe well the shape of the nuclear modification factor and interpret the maximum as being caused by the transverse flow picked up from the expanding medium. In addition, a large anisotropic elliptic flow (v_2) of NPE has been observed in Au+Au colisions at $\sqrt{s_{NN}}$= 200 GeV as demonstrated in Figure 7. The v_2 was measured by 2-particle correlation method and 4-particle cumulant method, where the latter one shows consistently smaller v_2 values as expected due to smaller contribution of non-flow effects. These v_2 values are systematically higher than those measured at lower collision energies $\sqrt{s_{NN}}$ = 39 and 62.4 GeV, respectively [47]. The large v_2 values provide additional indication of strong interaction of charm quarks with the medium and pose a challenge for model calculations to describe simultaneously both R_{AA} and v_2. However, future measurements of v_2 of charmed mesons are needed to obtain direct information about charm elliptic flow.

6. Outlook

In this contribution to these proceedings, only a selection of recent STAR results on correlations, jets, and heavy flavor production could be featured. Recent progress in studies of fully reconstructed jets in combination with high statistics Run-11 Au+Au data sample is expected to bring new insights into energy loss in medium created at top RHIC energy. From 2014 onward, the STAR experiment will run with greatly enhanced detection capabilities of heavy-flavor particles allowing direct reconstruction of secondary vertices of charm and beauty hadron decays. This together with the higher RHIC luminosity will enable STAR to enter a new era of precision and comprehensive measurements for rare probes of the nuclear matter.

Acknowledgements

The work has been supported by the grant 13-20841S of the Czech Science Foundation (GACR).

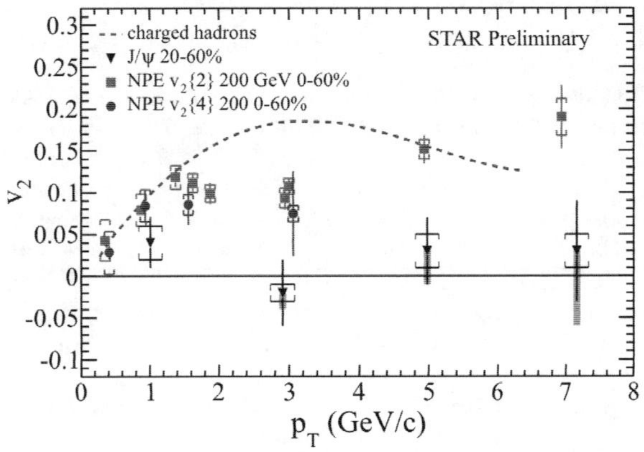

Figure 7. Measurement of heavy flavor electron (NPE) elliptic flow v_2 as a function of p_T in 0-60% Au+Au collisions at $\sqrt{s_{NN}}$ =200 GeV. For comparison, the charged hadron (dashed line) and J/ψ (closed triangle down) elliptic flow is also shown. The figure is taken from [47].

References

[1] Adams J *et al.* (STAR Collaboration) 2005 *Nucl.Phys.* **A757** 102–183 (*Preprint* nucl-ex/0501009)
[2] Adcox K *et al.* (PHENIX Collaboration) 2005 *Nucl.Phys.* **A757** 184–283 (*Preprint* nucl-ex/0410003)
[3] Back B *et al.* (PHOBOS Collaboration) 2005 *Nucl.Phys.* **A757** 28–101 (*Preprint* nucl-ex/0410022)
[4] Arsene I *et al.* (BRAHMS Collaboration) 2005 *Nucl.Phys.* **A757** 1–27 (*Preprint* nucl-ex/0410020)
[5] Evdokimov O *et al.* (STAR Collaboration) *these proceedings*
[6] Hamed A *et al.* (STAR Collaboration) *these proceedings*
[7] Gagliardi C *et al.* (STAR Collaboration) 2005 *Eur.Phys.J.* **C43** 263–270
[8] Jacobs P 2005 *Eur.Phys.J.* **C43** 467–473 (*Preprint* nucl-ex/0503022)
[9] Abelev B *et al.* (STAR Collaboration) 2009 *Phys.Rev.* **C80** 064912 (*Preprint* 0909.0191)
[10] Alver B *et al.* (PHOBOS Collaboration) 2010 *Phys.Rev.Lett.* **104** 062301 (*Preprint* 0903.2811)
[11] Alver B and Roland G 2010 *Phys.Rev.* **C81** 054905 (*Preprint* 1003.0194)
[12] Khachatryan V *et al.* (CMS Collaboration) 2010 *JHEP* **1009** 091 (*Preprint* 1009.4122)
[13] Aad G *et al.* (ATLAS Collaboration) 2012 *JHEP* **1205** 157 (*Preprint* 1203.3549)
[14] Abelev B *et al.* (ALICE Collaboration) 2013 *Phys.Lett.* **B719** 29–41 (*Preprint* 1212.2001)
[15] Chatrchyan S *et al.* (CMS Collaboration) 2013 *Phys.Lett.* **B718** 795–814 (*Preprint* 1210.5482)
[16] Aad G *et al.* (ATLAS Collaboration) 2013 *Phys.Rev.Lett.* **110** 182302 (*Preprint* 1212.5198)
[17] Arbuzov B, Boos E and Savrin V 2011 *Eur.Phys.J.* **C71** 1730 (*Preprint* 1104.1283)
[18] Dusling K and Venugopalan R 2013 *Phys.Rev.* **D87** 054014 (*Preprint* 1211.3701)
[19] Kovchegov Y V and Wertepny D E 2013 *Nucl.Phys.* **A906** 50–83 (*Preprint* 1212.1195)
[20] Wong C Y 2011 *Phys.Rev.* **C84** 024901 (*Preprint* 1105.5871)
[21] Strikman M 2011 *Acta Phys.Polon.* **B42** 2607–2630 (*Preprint* 1112.3834)
[22] Alderweireldt S and Van Mechelen P 2012 33–40 (*Preprint* 1203.2048)
[23] Werner K, Karpenko I and Pierog T 2011 *Phys.Rev.Lett.* **106** 122004 (*Preprint* 1011.0375)
[24] Avsar E, Flensburg C, Hatta Y, Ollitrault J Y and Ueda T 2011 *Phys.Lett.* **B702** 394–397 (*Preprint* 1009.5643)
[25] Deng W T, Xu Z and Greiner C 2012 *Phys.Lett.* **B711** 301–306 (*Preprint* 1112.0470)
[26] Bozek P and Broniowski W 2013 *Phys.Lett.* **B718** 1557–1561 (*Preprint* 1211.0845)
[27] Adare A *et al.* (PHENIX Collaboration) 2013 *Phys.Rev.Lett.* **111** 212301 (*Preprint* 1303.1794)
[28] Ajitanand N, Alexander J, Chung P, Holzmann W, Issah M *et al.* 2005 *Phys.Rev.* **C72** 011902 (*Preprint* nucl-ex/0501025)
[29] Wang F *et al.* (STAR Collaboration) *Hard Probes 2013 conference, proceedings to be published in Nucl.Phys.A*
[30] Agakishiev H *et al.* (STAR Collaboration) 2011 *Phys.Rev.* **C83** 061901 (*Preprint* 1102.2669)
[31] Adamczyk L *et al.* (STAR Collaboration) 2013 *Phys.Rev.* **C87** 044903 (*Preprint* 1212.1653)
[32] Renk T 2008 *Phys.Rev.* **C78** 014903 (*Preprint* 0804.1204)
[33] Cacciari M, Salam G P and Soyez G 2008 *JHEP* **0804** 063 (*Preprint* 0802.1189)

[34] Adamczyk L *et al.* (STAR Collaboration) 2013 (*Preprint* 1302.6184)
[35] Renk T 2013 *Phys.Rev.* **C87** 024905 (*Preprint* 1210.1330)
[36] Renk T 2011 *Phys.Rev.* **C84** 067902 (*Preprint* 1110.2313)
[37] Adamczyk L *et al.* (STAR Collaboration) 2012 *Phys.Rev.* **D86** 072013 (*Preprint* 1204.4244)
[38] Tlusty D (STAR collaboration) 2013 *Nucl.Phys.* **A904-905** 639c–642c (*Preprint* 1211.5995)
[39] Cacciari M, Nason P and Vogt R 2005 *Phys.Rev.Lett.* **95** 122001 (*Preprint* hep-ph/0502203)
[40] Dokshitzer Y L and Kharzeev D 2001 *Phys.Lett.* **B519** 199–206 (*Preprint* hep-ph/0106202)
[41] Adler S *et al.* (PHENIX Collaboration) 2005 *Phys.Rev.Lett.* **94** 082301 (*Preprint* nucl-ex/0409028)
[42] Adler S *et al.* (PHENIX Collaboration) 2006 *Phys.Rev.Lett.* **96** 032301 (*Preprint* nucl-ex/0510047)
[43] Abelev B *et al.* (STAR Collaboration) 2007 *Phys.Rev.Lett.* **98** 192301 (*Preprint* nucl-ex/0607012)
[44] Zhang B W, Wang E and Wang X N 2004 *Phys.Rev.Lett.* **93** 072301 (*Preprint* nucl-th/0309040)
[45] Wicks S, Horowitz W, Djordjevic M and Gyulassy M 2007 *Nucl.Phys.* **A784** 426–442 (*Preprint* nucl-th/0512076)
[46] Adil A and Vitev I 2007 *Phys.Lett.* **B649** 139–146 (*Preprint* hep-ph/0611109)
[47] Xie W (STAR Collaboration) 2013 *Nucl.Phys.* **A904-905** 170c–177c
[48] He M, Fries R J and Rapp R 2013 *Phys.Rev.Lett.* **110** 112301 (*Preprint* 1204.4442)
[49] Gossiaux P B, Aichelin J, Bluhm M, Gousset T, Nahrgang M *et al.* 2012 *PoS* **QNP2012** 160 (*Preprint* 1207.5445)

Journal of Physics: Conference Series **589** (2015) 012003 doi:10.1088/1742-6596/589/1/012003

Jet structure in pp and Pb-Pb collisions with the ALICE experiment

Oliver Busch for the ALICE collaboration

Universität Heidelberg, Physikalisches Institut, Im Neuenheimer Feld 226, 69120 Heidelberg, Germany

E-mail: o.busch@gsi.de

Abstract.
We report results from the ALICE collaboration on jet production and jet properties in pp and Pb-Pb collisions at the LHC. The charged particle jet production cross-section and measurements of jet fragmentation and jet shape in pp collisions at $\sqrt{s}$ = 7 TeV are presented. The results are confronted with simulations from Monte Carlo event generators. Measurements of charged particle jets in Pb-Pb collisions at $\sqrt{s_{NN}}$ = 2.76 TeV with resolution parameter $R = 0.2$ and $R = 0.3$ show a strong momentum dependent suppression in central relative to peripheral collisions. The analysis of the semi-inclusive distribution of charged particle jets recoiling from a high-p_{T} trigger hadron allows an unbiased measurement of the jet structure for larger cone radii. No significant redistribution of the energy flow inside the jet cone is observed. Di-hadron correlations allow to assess jet structure at very low particle momenta. For $2 < p_{T,trig} < 3$ GeV/c and $1 < p_{T,assoc} < 2$ GeV/c, a strongly asymmetric near-side correlation, broader in $\Delta\eta$ than $\Delta\phi$, is observed.

1. Introduction

Jets are collimated sprays of particles associated with hard-scattered partons. The study of jet production and fragmentation allows us to test our understanding of perturbative and non-perturbative aspects of QCD. In heavy-ion collisions, jets produced in the initial stage probe the early hot and dense quark gluon plasma phase of the fireball evolution. Interactions with the medium result in collisional energy loss and give rise to additional induced radiation, which emitted at small angles inside the jet cone can lead to a broadening of the jet profile.

The coupling between partons and medium at a scale of the order of the initial parton energy may be weak, governed by perturbative dynamics. At a scale of the order of ~1 GeV, set by the medium temperature or the saturation scale, the coupling between the jet and the medium may be strong [1]. Empirical observations suggest that indeed the thermal scale is relevant for the parton energy loss [2]. It has been predicted, that the coupling of jet fragments to the longitudinally flowing medium [3] or to turbulent color fields [4] will lead to broadening that is larger in pseudorapidity η than azimuthal angle ϕ (excentric jets). Jet reconstruction and measurements of the jet properties aim to quantify the in-medium jet energy loss and determine the dynamics of jet quenching. At low particle momenta, where full jet reconstruction is difficult due to strong underlying event backgrounds, the jet structure can be assessed via di-hadron correlation measurements.

Journal of Physics: Conference Series **589** (2015) 012003 doi:10.1088/1742-6596/589/1/012003

2. Results from pp collisions

Charged jets are reconstructed in the ALICE [5] central barrel from primary charged particle tracks measured in the Inner Tracking System (ITS) and the Time Projection Chamber (TPC). Tracks with transverse momentum $p_{\mathrm{T}} > 150$ MeV/c in the pseudo-rapidity interval $|\eta| < 0.9$ are clustered with the FastJet [6] anti-k_{T} algorithm with a resolution parameter $R = 0.4$ using the boost-invariant p_{T} recombination scheme. Jets which are fully contained within the detector acceptance are selected. The jet shape and jet constituents' transverse momentum spectra (jet fragmentation) are measured with charged particles in the leading (highest p_{T}) jet in each event. The jet spectra, shape, and fragmentation distributions are corrected for detector effects via unfolding or bin-by-bin corrections based on detector simulations.

In the top left panel of Fig. 1, the charged jet cross section [9] at $\sqrt{s} = 7$ TeV is shown. The top right panel presents the charged jet fragment distribution in scaled transverse momentum $\xi = \ln(p_{\mathrm{T}}^{jet,ch}/p_{\mathrm{T}}^{particle})$ [10]. The wide ALICE track momentum coverage down to $p_{\mathrm{T}} = 150$ MeV/c allows us to measure the distribution over more than 5 units in ξ.

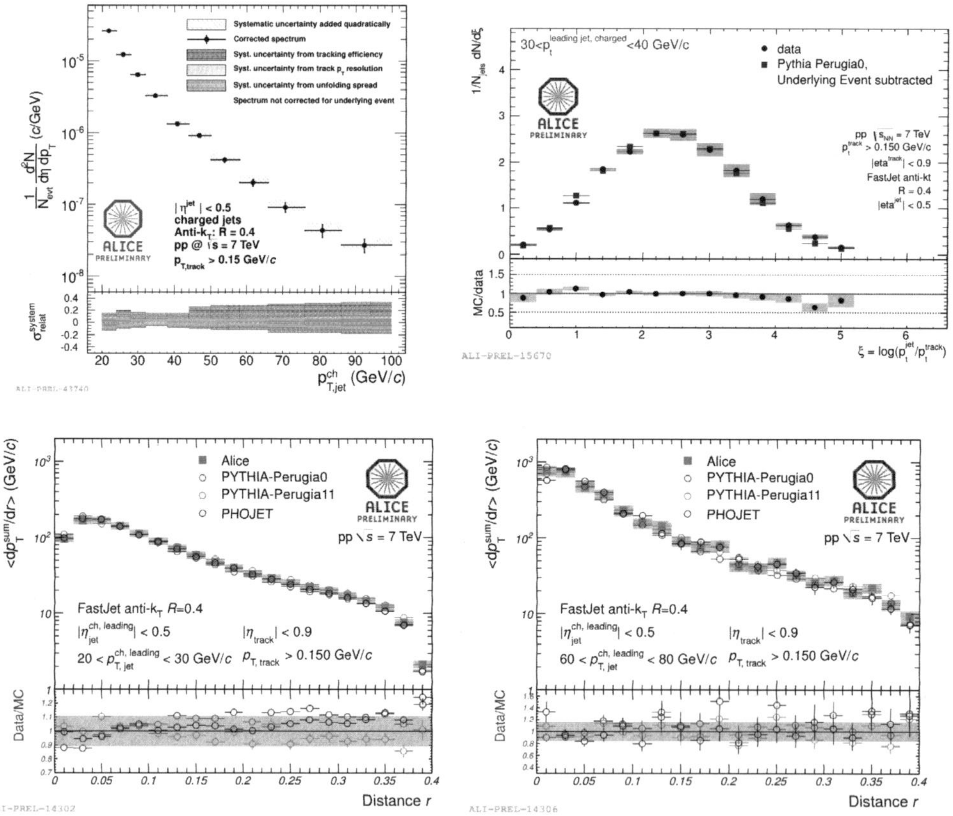

Figure 1. Results from pp collisions at $\sqrt{s} = 7$ TeV. Top left: charged jet cross section for $R = 0.4$. Systematic uncertainties are indicated by the boxes, the lower panel shows the contributions from detector effects and unfolding superimposed linearly. Top right: scaled transverse momentum distribution of jet fragments measured in leading charged jets with $30 < p_{\mathrm{T,jet}}^{\mathrm{ch}} < 40$ GeV/c. Bottom: radial distributions of transverse momentum density for $20 < p_{\mathrm{T,jet}}^{\mathrm{ch}} < 30$ GeV/c (left) and $60 < p_{\mathrm{T,jet}}^{\mathrm{ch}} < 80$ GeV/c (right). All results are corrected for detector effects and underlying event (see text). Jet shape and fragmentation distributions are compared to calculations from MC event generators.

The distribution exhibits the characteristic hump-backed plateau structure indicating QCD coherence [11]. The area of the distribution is equal to the charged particle multiplicity of the fragments. The bottom panels show the radial distributions of transverse momentum density around the jet axis [12]. The slope of the distribution measures the jet collimation. The increase in slope from $20 < p^{\mathrm{ch}}_{\mathrm{T,jet}} < 30$ GeV/c (bottom left) to $60 < p^{\mathrm{ch}}_{\mathrm{T,jet}} < 80$ GeV/c (bottom right) at small r indicates stronger collimation with increasing jet p_{T}.

The results are compared to Monte Carlo (MC) simulations from event generators: PYTHIA [13] tune Perugia0 [14] (fragmentation, jet shapes), PYTHIA tune Perugia2011 and PHOJET [15] (jet shapes). The underlying event contribution was measured inside perpendicular cones transverse to the jet axis and subtracted from the jet cross section, fragmentation and shape measurements, consistently in data and simulations. The data are reasonably well described by the simulations.

3. Jets in Pb-Pb collisions

Jet reconstruction in heavy-ion collisions is faced with a large background from the underlying event uncorrelated to hard parton scattering. The background contribution is reduced using small values of the jet resolution parameter of $R = 0.2$ and $R = 0.3$. The average charged background energy density is calculated event by event as the median p_{T} density of FastJet k_{T} clusters and subtracted jet by jet. The spectra are corrected for detector effects and background

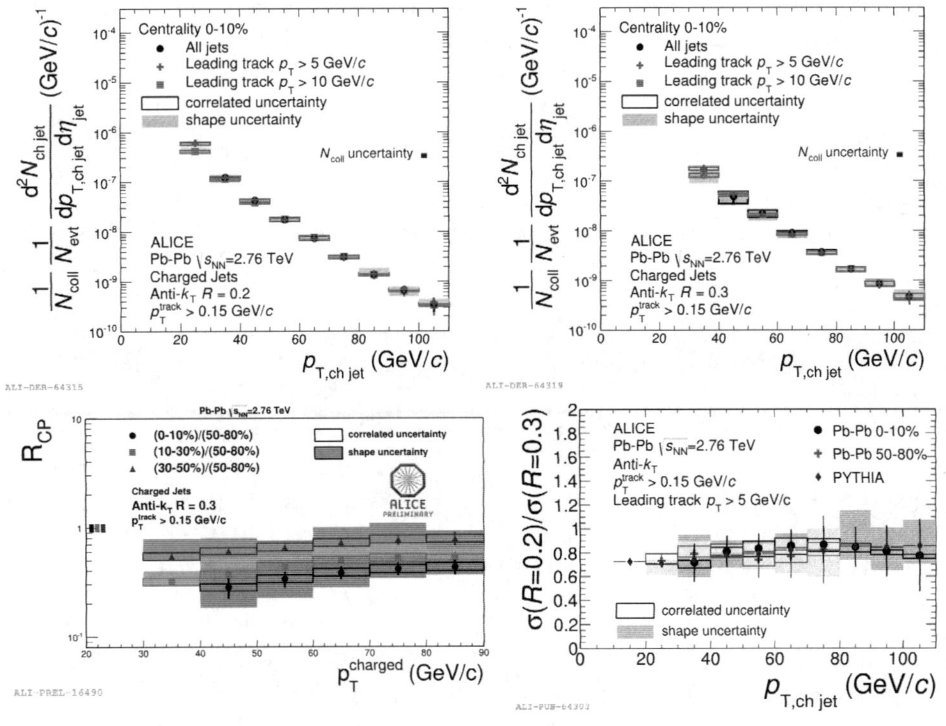

Figure 2. Single inclusive charged jet measurements for 10% most central Pb-Pb collisions at $\sqrt{s_{NN}} = 2.76$ TeV. Top: jet spectrum for $R = 0.2$ (left) and $R = 0.3$ (right) for different leading track selections, normalized to the number of binary collisions. Correlated (shape) systematic uncertainties are indicated by the boxes (shaded bands). Bottom left: jet nuclear modification factor R_{CP} for different centrality selections. Bottom right: spectral ratio for different R compared to PYTHIA calculations.

Journal of Physics: Conference Series **589** (2015) 012003 doi:10.1088/1742-6596/589/1/012003

fluctuations via unfolding. Background fluctuations also give rise to low p_{T} 'combinatorial jets' composed of background particles. Their contribution can be suppressed, for example, by introducing a fragmentation bias requiring a high momentum jet constituent.

Fig. 2 shows single charged jet spectra in Pb-Pb collisions at $\sqrt{s_{NN}} = 2.76$ TeV for $R = 0.2$ (top left) and $R = 0.3$ (top right) [16, 17, 18]. The spectra are normalized to the number of binary collisions. The inclusive jet spectra are compared to results requiring at least one jet constituent with $p_{\mathrm{T}} > 5$ or 10 GeV/c. Within uncertainties, no dependence on the momentum of the leading constituent is observed, which indicates that the longitudinal momentum distribution of high-p_{T} tracks of jets reconstructed in Pb-Pb collisions remains largely unmodified.

The centrality dependence of the jet quenching can be quantified using $R_{\mathrm{CP}}(p_{\mathrm{T}}) = \frac{\langle N_{\mathrm{coll}}^{P}\rangle}{\langle N_{\mathrm{coll}}^{C}\rangle} \frac{\mathrm{d}^2 N_{\mathrm{jet}}^{C}/\mathrm{d}p_{\mathrm{T,jet}}\,\mathrm{d}\eta}{\mathrm{d}^2 N_{\mathrm{jet}}^{\mathrm{P}}/\mathrm{d}p_{\mathrm{T,jet}}\,\mathrm{d}\eta}$, where C stands for central and P for peripheral. The measured R_{CP} (bottom left) is significantly smaller than unity. The suppression is strongest at low jet momentum, and at high momentum R_{CP} rises to values of approximately 0.4. In the bottom right panel, the ratio of the charged jet spectra obtained for $R = 0.2$ and $R = 0.3$ is shown for two centrality classes. This ratio is sensitive to the jet structure. The data are consistent with the PYTHIA values for peripheral and central events: no broadening of the hard core of the reconstructed jets is observed within the present systematic uncertainties.

4. Hadron triggered recoil jets in Pb-Pb collisions

The analysis of hadron triggered recoil jets allows us to extend the study of jet quenching to larger values of R. The semi-inclusive distribution of charged jets recoiling back-to-back from a high-p_{T} charged hadron in Pb-Pb collisions [19] is measured. Model studies [20] show that a high p_{T} hadron trigger introduces a geometrical bias, towards jets generated on the surface of the fireball and directed outward. Jets recoiling from such a trigger tend to have a larger in-medium path length. The recoil jet p_{T} distribution normalized per trigger is shown in the upper left panel of Fig. 3 for different hadron trigger intervals. After subtraction of the average background, a large fraction of the jet population is assigned a negative energy. This part of the spectrum is dominated by combinatorial jets, and is uncorrelated with the trigger p_{T}. For high $p_{\mathrm{T,jet}}$, the recoil jet distribution is seen to depend strongly on p_T^{trig}, since a harder hadron trigger selects on average a larger Q^2.

In the difference of the spectra for different intervals of trigger hadron p_{T}, the contribution of combinatorial jets uncorrelated to the hadron trigger cancels:

$$\Delta_{recoil}((p_{\mathrm{T}}^{\mathrm{trig},1} - p_{\mathrm{T}}^{\mathrm{trig},2}) - (p_{\mathrm{T}}^{\mathrm{trig},3} - p_{\mathrm{T}}^{\mathrm{trig},4})) = \frac{1}{N_{\mathrm{trig}}} \frac{\mathrm{d}N}{\mathrm{d}p_{\mathrm{T,jet}}^{\mathrm{ch}}} (p_{\mathrm{T}}^{\mathrm{trig},1} < p_{\mathrm{T}}^{\mathrm{trig}} < p_{\mathrm{T}}^{\mathrm{trig},2}) - c \frac{1}{N_{\mathrm{trig}}} \frac{\mathrm{d}N}{\mathrm{d}p_{\mathrm{T,jet}}^{\mathrm{ch}}} (p_{\mathrm{T}}^{\mathrm{trig},3} < p_{\mathrm{T}}^{\mathrm{trig}} < p_{\mathrm{T}}^{\mathrm{trig},4}) \quad (1)$$

The Δ_{recoil} spectrum for $R = 0.4$ is shown in the upper right panel of Fig. 3 for hadron trigger intervals from 20-50 GeV/c and 15-20 GeV/c, using a scale factor $c = 0.956$. This novel variable allows us to study an unbiased sample of jets reconstructed with a large radius, maintaining a jet constituent cutoff as low as $p_{\mathrm{T}}^{\mathrm{const}} > 0.15$ GeV/c.

To explore the energy redistribution within recoil jets, we consider the ratio for the measured Δ_{recoil} distribution over PYTHIA (tune Perugia2010) simulations ΔI_{AA}^{PYTHIA}, presented in the bottom panels of Fig. 3. We find similar values of ΔI_{AA}^{PYTHIA} for $R = 0.2$ (left) and $R = 0.4$ (right) and constituent p_{T} threshold [19] (not shown). Within present experimental uncertainties, we do not observe significant redistribution of the jet energy.

5. Jet-like near-side peak shapes in Pb-Pb collisions

In the analysis of di-hadron correlations, the angular differences $\Delta\eta$ and $\Delta\phi$ between trigger particles and all associated particles satisfying pairs of cuts on $(p_{\mathrm{T,trig}}, p_{\mathrm{T,assoc}})$ are studied. We use tracks reconstructed in the TPC within $|\Delta\eta| < 0.9$ constrained to the primary vertex. The

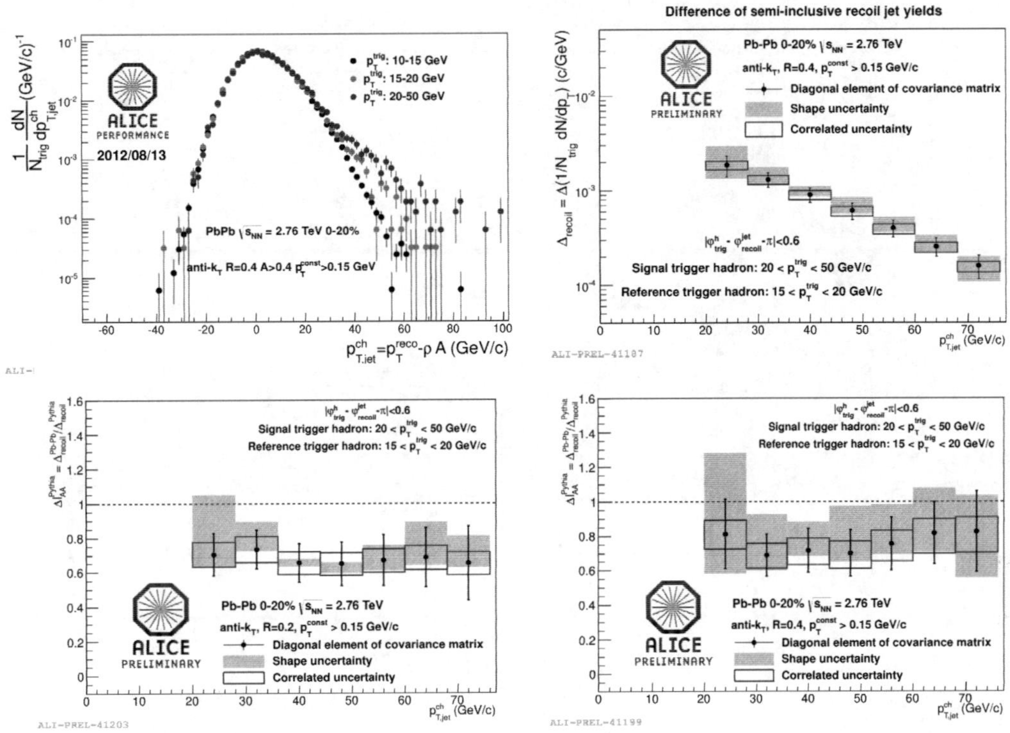

Figure 3. Upper left: Semi-inclusive recoil jet spectrum for 0-20% central Pb-Pb collisions at $\sqrt{s_{NN}} = 2.76$ TeV, for different trigger p_T classes. Jets reconstructed with the anti-k_T algorithm for $R = 0.4$, $p_T^{const} > 0.15$ GeV/c. Upper right: Semi-inclusive recoil jet spectrum difference Δ_{recoil}. Lower panels: ΔI_{AA} for recoil jets using PYTHIA as a reference, for $R = 0.2$ (left) and $R = 0.4$ (right).

angular correlations are quantified by the per trigger associated yield $1/N_{trig}\ \mathrm{d}^2N/\mathrm{d}\Delta\phi \mathrm{d}\Delta\eta$. The uncorrected distributions at the near-side ($\Delta\phi = \Delta\eta = 0$) represent the sum of a jet-like correlation 'signal' and a 'background' induced by $\Delta\eta$ independent long-range correlations (mostly due to hydrodynamic flow [7] related to the initial spatial anisotropy of the overlap zone of the two colliding nuclei). This background is estimated by introducing a gap in pseudo-rapidity $1 < |\Delta\eta| < 1.6$, and subtracted. The associated yield is corrected for acceptance, detector effects and secondary particle contamination [8].

The near-side di-hadron correlations are studied for different ($p_{\mathrm{T,trig}}, p_{\mathrm{T,assoc}}$) ranges. In the top left panel of Fig. 4, the per trigger associated yield for the lowest p_T bin ($2 < p_{\mathrm{T,trig}} < 3$ GeV/c, $1 < p_{\mathrm{T,assoc}} < 2$ GeV/c) for the 10% most central collisions is shown. The distribution is asymmetric, and a strong broadening in $\Delta\eta$ is observed. The projections on $\Delta\eta$ and $\Delta\phi$ (upper right) indicate a flattening in the region $|\Delta\eta| < 0.4$. The centrality and p_T dependence of the rms $\sigma_{\Delta\phi}$, $\sigma_{\Delta\eta}$ of the distributions is presented in the bottom panels of Fig. 4. The near-side peak width decreases with increasing ($p_{\mathrm{T,trig}}, p_{\mathrm{T,assoc}}$), reflecting the stronger collimation of higher p_T jets. Within uncertainties, no centrality dependence of $\sigma_{\Delta\phi}$ is observed, whereas $\sigma_{\Delta\eta}$ increases from the pp reference (centrality 100% data points) to the most central Pb-Pb collisions.

The data points are compared with the result of an AMPT 2.25 (with string melting) MC simulation [21] shown as lines. The pp values have been simulated with PYTHIA (tune Perugia0 [14]). AMPT simulates the initial conditions using HIJING [22]. It includes re-scattering of

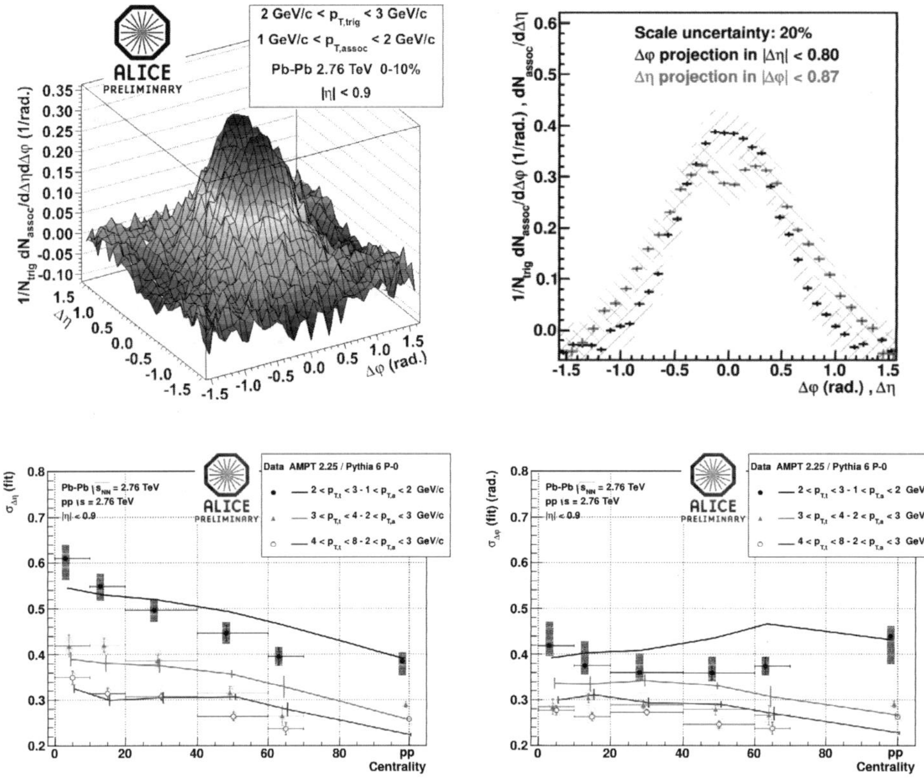

Figure 4. Upper left: Per trigger associated yield differential in $\Delta\eta$ and $\Delta\phi$ for the 10% most central Pb-Pb collisions at $\sqrt{s_{NN}} = 2.76$ TeV, and the lowest p_T bin studied. Upper right: Corresponding projections. Bottom: centrality dependence of $\sigma_{\Delta\eta}$ (left) and $\sigma_{\Delta\phi}$ (right) for three different bins of $p_{T,\mathrm{trig}}$ and $p_{T,\mathrm{assoc}}$. The lines indicate results of AMPT and PYTHIA calculations.

partons and hadrons. Hadronization is simulated using the Lund model and coalescence. The AMPT simulations are in good agreement with our measurements.

References

[1] Müller B 2011 *Nucl. Phys.* A **855** 74
[2] Renk T 2013 *Nucl. Phys.* A **904** 725
[3] Armesto N *et al.* 2004 *Phys. Rev. Lett.* **93** 242301
[4] Majumder A *et al.* 2007 *Phys. Rev. Lett.* **99** 042301
[5] Aamodt K *et al.* 2008 *J. Instrum.* **3** S08002
[6] Cacciari M, Salam G P and Soyez G 2012 *EPJ* **72** 1896
[7] Poskanzer A M, Voloshin S 1998 *Phys. Rev.* C **58** 1671
[8] Morsch A [ALICE collaboration] 2013 *Nucl. Phys.* A **910** 281
[9] Vajzer M [ALICE Collaboration] 2012 *arXiv:1212.6890 [nucl-ex]*
[10] Ma R [ALICE collaboration] 2013 *Nucl. Phys.* A **910** 319
[11] Ermolayev B I, Fadin V S 1981 *JETP Lett.* **33** 285; Mueller A H 1981 *Phys. Lett.* B **104** 161
[12] Prasad S K [ALICE Collaboration] 2012 *J. Phys. Conf. Ser.* **389** 012005
[13] Sjostrand T, Mrenna S and Skands P Z 2006 *JHEP* **05** 026
[14] Skands P Z 2010 *Phys. Rev.* D **82** 074018
[15] Roesler S, Engel R and Ranft J 2000 *Advanced Monte Carlo, Lisbon* 1033, *arXiv:0012252 [hep-ph]*.
[16] Verweij M [ALICE Collaboration] 2013 *Nucl. Phys.* A **910** 421
[17] Verweij M [ALICE Collaboration] 2013 *Nucl. Phys.* A **904** 1015

[18] ALICE Collaboration *arXiv:1311.0633 [nucl-ex]*.
[19] Cunqueiro L [ALICE Collaboration] 2013 *Nucl. Phys.* A **904** 728
[20] Renk T 2012 *Phys. Rev.* C **87** 024905
[21] Lin Z W *et al.* 2005 *Phys. Rev.* C **72** 064901, Xu J and Ko C M 2011 *Phys. Rev.* C **83** 034904
[22] Wang X N and Gyulassi M 1991 *Phys. Rev.* D **44** 3501

Journal of Physics: Conference Series **589** (2015) 012004 doi:10.1088/1742-6596/589/1/012004

Radiation spectrum of a massive quark-gluon antenna

M R Moldes[12], M R Calvo[13] and C A Salgado[1]

[1]Departamento de Física de Partículas and IGFAE, Universidade de Santiago de Compostela, E-15706 Santiago de Compostela, Galicia, Spain

[2]Centre de Physique Théorique, École Polytechnique, CNRS, 91128 Palaiseau, France

[3]Institut de Physique Theorique de Saclay, F-91191 Gif-sur-Yvette, France

E-mail:
manoel.rodriguez-moldes@usc.es,manoel.rodriguez@usc.es,carlos.salgado@usc.es

Abstract.
We compute the radiation spectrum of antennas containing a heavy quark, focusing on the case of a massive quark-gluon antenna. We apply the classical Yang-Mills formalism (CYM), and treat the interactions of the jet with the medium in the multiple soft scattering approximation. This work completes the studies of antenna radiation inside a medium. The main conclusion is that decorrelation occurs faster in the case in which at least one of the emitters is a heavy quark, and is faster in the massive quark-gluon case.

1. Introduction

The hot and dense state of matter created in heavy ion collisions is known to modify the evolution of jets. This has been experimentally studied first at the RHIC at BNL [1] and then at the LHC at CERN [2, 3, 4]. The most successful model for this jet quenching effect is the one based on the enhancement of the gluon radiation spectrum induced by the medium [5, 6, 7, 8, 9, 10, 11, 12].

In this model, radiation off massive quarks is expected to be suppressed in regions of phase space relevant for phenomenological applications due to the dead-cone effect. On the other hand, RHIC data on the suppression of the non-photonic electrons (expected to be dominated by heavy quark decays) is compatible, taking it at face value, with no mass effect in the radiation [13]. In order to understand this experimental data, it seems that new developments in the theoretical description are needed.

Several improvements have been made in the theoretical side in the recent years[14, 15, 16, 17]. Also, efforts have been made so as to compute interference effects, which are known to play a significant role in the vacuum case [18, 19, 20, 21, 22], leading to the well known fact that subsequent emissions are angular ordered. The emerging picture in the medium case is that interferences become important when the transverse size of the antenna is smaller compared to the characteristic transverse size of the medium, and are irrelevant in the opposite case [23, 24, 25, 27, 28, 29].

Motivated by all the above considerations, we compute in this work the interference effects for antennas containing heavy quarks. We perform the calculation using a semiclassical approach and treat the interactions of the partons with the medium in the multiple soft scattering

Journal of Physics: Conference Series **589** (2015) 012004 doi:10.1088/1742-6596/589/1/012004

approximation. The main conclusion of the paper is that coherence is more easily lost when one of the emitters is a heavy quark, and that a massive quark-gluon antenna loses coherence faster than a massive quark-antiquark antenna. The loss of coherence is an effect that in general implies a larger energy loss.

2. Calculation of the amplitude

The amplitude for one gluon emission can be calculated using the reduction formula [25]

$$\mathcal{M}^a(k) = -\sum_\lambda \int_{x^+=+\infty} dx^- d^2\mathbf{x}\, e^{ik\cdot x}\, 2\partial_x^+ \mathbf{A}^a(x)\cdot \boldsymbol{\epsilon}_\lambda(\vec{k}) \tag{1}$$

with $k^\mu = (\omega, \vec{k})$ being the 4-momentum of the emitted gluon and $\mathbf{A}$ the transverse gauge field. The gauge field is obtained from the classical Yang-Mills (CYM) equations

$$[D_\mu, F^{\mu\nu}] = J^\nu \tag{2}$$

where $D_\mu \equiv \partial_\mu - igA_\mu$ and $F_{\mu\nu} \equiv \partial_\mu A_\nu - \partial_\nu A_\mu - ig[A_\mu, A_\nu]$, and with the current J^μ being covariantly conserved, i.e.,

$$[D_\mu, J^\mu] = 0 \tag{3}$$

The initial state of a component of the antenna with momentum $p^\mu = (E, \vec{p})$ and charge color vector Q^a is represented by this vacuum current:

$$J^{\mu,a}_{(0)}(x) = g\frac{p^\mu}{E}\delta^{(3)}\left(\vec{x} - \frac{\vec{p}}{E}t\right)\theta(t)Q^a \tag{4}$$

The medium will affect this vacuum current by inducing a color rotation:

$$J^\mu(x) = U_p(x^+, 0)J^\mu_{(0)}(x) \tag{5}$$

described by a Wilson line:

$$U_p(x^+, 0; \mathbf{r}) \equiv \mathcal{P}\exp\left\{\int_0^{x^+} d\xi\, T\cdot A^-_{\text{med}}(\xi, \mathbf{p}\xi/p^+)\right\} \tag{6}$$

Leaving only terms that are linear on the medium induced field and performing the calculation in the light-cone gauge ($A^+ = 0$, with light-cone coordinates defined as $A^\pm = (A^0 \pm A^3)/\sqrt{2}$ and $\mathbf{A} = (A^1, A^2)$), we get the following expression for the amplitude:

$$\begin{aligned}\mathcal{M}^a(\vec{k}) = \sum_\lambda \frac{g}{k^+} \int_{x^+=+\infty} d^2\mathbf{x}\, e^{ik^-x^+} e^{-i\mathbf{k}\cdot\mathbf{x}} \int_0^{+\infty} dy^+\, e^{ik^+\frac{p^-}{p^+}y^+} \\ \times\, \boldsymbol{\epsilon}_\lambda(k)\cdot(i\boldsymbol{\partial}_y + k^+\mathbf{n})\,\mathcal{G}^{ab}(x^+, \mathbf{x}; y^+, \mathbf{y}|k^+)\Big|_{\mathbf{y}=\mathbf{n}y^+} U_p^{bc}(y^+, 0)Q^c\end{aligned} \tag{7}$$

where we have explicited the color structure, defined the dimensionless vector $\mathbf{n} = \mathbf{p}/p^+$ and $\mathcal{G}$ is a Green's function that takes into account both the color rotation of the gluon and its Brownian motion in the transverse plane due to interactions with the medium field. These features of the Green's function $\mathcal{G}$ can be easily seen thanks to its expression as a path integral in the transverse plane:

$$\mathcal{G}(x^+, \mathbf{x}; y^+, \mathbf{y}|k^+) = \int_{\mathbf{r}(y^+)=\mathbf{y}}^{\mathbf{r}(x^+)=\mathbf{x}} \mathcal{D}\mathbf{r}\exp\left\{\frac{ik^+}{2}\int_{y^+}^{x^+} d\xi\, \dot{\mathbf{r}}^2(\xi)\right\} U(x^+, y^+; \mathbf{r}) \tag{8}$$

Reached this point, we wonder what could be the effect due to the presence of a mass M. To see that, we recall that the only difference between a massive quark and a massless one comes from the dispersion relation:

$$2p^+p^- - \mathbf{p}^2 = M^2 \tag{9}$$

So we can understand the existence of a mass as a modification to the relation of p^- with respect of the other components of the momentum. Since the only place where this p^- component appears (provided we neglect any dependence of the medium field on the x^- variable) in the previous expressions is in an exponential, the presence of mass brings us a new phase.

$$\exp\left(ik^+\frac{p^-}{p^+}y^+\right) = \exp\left(i\frac{k^+}{2}\theta_{DC}^2 y^+\right)\exp\left(i\frac{k^+}{2}\mathbf{n}^2 y^+\right) \tag{10}$$

where θ_{DC} is the so called *dead-cone angle*, defined as

$$\theta_{DC} \equiv \frac{M}{p^+} \tag{11}$$

From this point on, we denote with a bar quantities (such as momentum) related to the gluon leg of the antenna, while those related to the quark leg are left without bar; e.g., p^μ represents the quark momentum and $\bar{p}^\mu$ represents the gluon momentum.

Writing down the two possible amplitudes, i.e.,

- when the gluon is radiated off a heavy quark:

$$\begin{aligned}\mathcal{M}_q^a(\vec{k}) = \sum_\lambda \frac{g}{k^+} \int\limits_{x^+=+\infty} d^2\mathbf{x}\, e^{ik^-x^+} e^{-i\mathbf{k}\cdot\mathbf{x}} \int_0^{+\infty} dy^+ \exp\left[i\frac{k^+}{2}(\theta_{DC}^2 + \mathbf{n}^2)y^+\right] \\ \times\, \boldsymbol{\epsilon}_\lambda(k)\cdot(i\boldsymbol{\partial}_y + k^+\mathbf{n})\,\mathcal{G}^{ab}(x^+,\mathbf{x};y^+,\mathbf{y}|k^+)\Big|_{\mathbf{y}=\mathbf{n}y^+} U_p^{bc}(y^+,0)Q_q^c\end{aligned} \tag{12}$$

- when the gluon is radiated off a gluon:

$$\begin{aligned}\mathcal{M}_g^a(\vec{k}) = \sum_\lambda \frac{g}{k^+} \int\limits_{x^+=+\infty} d^2\mathbf{x}\, e^{ik^-x^+} e^{-i\mathbf{k}\cdot\mathbf{x}} \int_0^{+\infty} dy^+ \exp\left[i\frac{k^+}{2}\bar{\mathbf{n}}^2 y^+\right] \\ \times\, \boldsymbol{\epsilon}_\lambda(k)\cdot(i\boldsymbol{\partial}_y + k^+\bar{\mathbf{n}})\,\mathcal{G}^{ab}(x^+,\mathbf{x};y^+,\mathbf{y}|k^+)\Big|_{\mathbf{y}=\bar{\mathbf{n}}y^+} U_{\bar{p}}^{bc}(y^+,0)Q_g^c\end{aligned} \tag{13}$$

we can explicitly see the difference due to the presence of mass: a complex phase in (12) that (13) lacks.

3. Calculation of the spectrum

With the amplitudes (12) and (13) we can compute the radiation spectrum of the antenna, which reads:

$$dN = \frac{\alpha_s}{(2\pi)^2}\left[C_F\mathcal{R}_q + C_A\mathcal{R}_g - C_A\mathcal{J}\right]\frac{d^3k}{(k^+)^3} \tag{14}$$

where we have defined the independent radiation off the heavy quark $\mathcal{R}_q$

$$C_F\mathcal{R}_q = (k^+)^2\langle|\mathcal{M}_q|^2\rangle \tag{15}$$

the independent radiation off the gluon $\mathcal{R}_g$

$$C_A\mathcal{R}_g = (k^+)^2\langle|\mathcal{M}_g|^2\rangle \tag{16}$$

and the interference spectrum between both emitters $\mathcal{J}$

$$-C_A\mathcal{J} = (k^+)^2\,\mathrm{Re}\langle\mathcal{M}_q\mathcal{M}_g^\dagger\rangle \tag{17}$$

Journal of Physics: Conference Series **589** (2015) 012004 doi:10.1088/1742-6596/589/1/012004

3.1. Independent Radiation Spectrum.

The spectrum of independent radiation off the heavy quark $\mathcal{R}_q$ is directly evaluated from (15) taking the limit $\mathbf{n} \to 0$ and it reads

$$\mathcal{R}_q = 2\mathrm{Re} \int_0^\infty \mathrm{d}y'^+ \int_0^{y'^+} \mathrm{d}y^+ \exp\left[i\frac{k^+}{2}\theta_{DC}^2(y^+ - y'^+)\right] \times \\ \times \int \mathrm{d}^2\mathbf{z}\, \exp\left[-i\mathbf{k}\cdot\mathbf{z} - \frac{1}{2}\int_{y'^+}^\infty \mathrm{d}\xi\, n(\xi)\sigma(\mathbf{z})\right] \boldsymbol{\partial}_y \cdot \boldsymbol{\partial}_z \mathcal{K}(y'^+, \mathbf{z}; y^+, \mathbf{y})\big|_{\mathbf{y}=\mathbf{0}} \quad (18)$$

This expression serves as a check of the formalism, since it is the same found in [26].

For the independent radiation off the gluon we can evaluate (16) or, given the similarities between (12) and (13), just take the limit $\theta_{DC} \to 0$ in (18), which gives us the following result:

$$\mathcal{R}_g = 2\mathrm{Re} \int_0^\infty \mathrm{d}y'^+ \int_0^{y'^+} \mathrm{d}y^+ \int \mathrm{d}^2\mathbf{z}\, \exp\left[-i\mathbf{k}\cdot\mathbf{z} - \frac{1}{2}\int_{y'^+}^\infty \mathrm{d}\xi\, n(\xi)\sigma(\mathbf{z})\right] \times \\ \times \boldsymbol{\partial}_y \cdot \boldsymbol{\partial}_z \mathcal{K}(y'^+, \mathbf{z}; y^+, \mathbf{y})\big|_{\mathbf{y}=\mathbf{0}} \quad (19)$$

3.2. Interference Spectrum.

Taking the medium averages and following the same procedure as in [25], we found that the interference spectrum $\mathcal{J}$ yields

$$\mathcal{J} = \mathrm{Re}\Bigg\{ \int_0^\infty \mathrm{d}y'^+ \int_0^{y'^+} \mathrm{d}y^+ \left(1 - \Delta(y^+, 0)\right) \exp\left[i\frac{k^+}{2}(\theta_{DC}^2 + \delta\mathbf{n}^2)y^+\right] \times \\ \times \int \mathrm{d}^2\mathbf{z} \exp\left[-i\bar{\boldsymbol{\kappa}}\cdot\mathbf{z} - \frac{1}{2}\int_{y'^+}^\infty \mathrm{d}\xi\, n(\xi)\sigma(\mathbf{z})\right] \times \\ \times (\boldsymbol{\partial}_y - ik^+\delta\mathbf{n}) \cdot \boldsymbol{\partial}_z \mathcal{K}(y'^+, \mathbf{z}; y^+, \mathbf{y})\big|_{\mathbf{y}=\delta\mathbf{n}y^+} \Bigg\} + \mathrm{sym.} \quad (20)$$

with $\bar{\boldsymbol{\kappa}} = \mathbf{k} - k^+\bar{\mathbf{n}}$ (and $\boldsymbol{\kappa} = \mathbf{k} - k^+\mathbf{n}$) and $\mathcal{K}$ being the path integral

$$\mathcal{K}(y'^+, \mathbf{z}; y^+, \mathbf{y}|k^+) = \int_{\mathbf{r}(y^+)=\mathbf{y}}^{\mathbf{r}(y'^+)=\mathbf{z}} \mathcal{D}\mathbf{r} \exp\left\{\int_{y^+}^{y'^+} \mathrm{d}\xi \left(i\frac{k^+}{2}\dot{\mathbf{r}}^2(\xi) - \frac{1}{2}n(\xi)\sigma(\mathbf{r})\right)\right\} \quad (21)$$

that takes into account the gluon multiple scattering with the medium and its Brownian motion in the transverse plane from $\mathbf{r}(y^+) = \mathbf{y}$ to $\mathbf{r}(y'^+) = \mathbf{z}$. The symmetric part is obtained interchanging $q \leftrightarrow g$, i.e., putting a bar in quantities without it and viceversa.

The only place where mass appears in the interference spectrum (20) is in the phase $\exp(ik^+\theta_{DC}^2 y^+/2)$. If the dead-cone angle is large, satisfying

$$\theta_{DC} \gg \sqrt{\frac{2}{k^+L}} \quad (22)$$

where L is the medium length, the mass phase will oscillate quickly, annihilating the interference. Putting it in space coordinates, this decoherence condition reads

$$\theta_{DC} \gg \frac{1}{\sqrt{\omega L}} \quad (23)$$

This loss of coherence implies a larger energy loss.

3.3. *Comparison with a massive quark-antiquark antenna*

We can also compute the interference spectrum for a massive quark-antiquark antenna. In this case, the mass phase is:

$$\exp\left[i\frac{k^+}{2}(\theta_{DC}^2 y^+ - \bar{\theta}_{DC}^2 y'^+)\right]$$

while in the massive quark-gluon case it was:

$$\exp\left(i\frac{k^+}{2}\theta_{DC}^2 y^+\right)$$

In other words, the mass phase for a heavy quark-antiquark antenna goes with the gluon time formation, while for a heavy quark-gluon antenna it goes with the length of the medium. Thus, the interference between emitters will be destroyed faster in a heavy qg antenna.

4. Summary

In this work the interference effects for antennas containing massive quarks are calculated in a semiclassical framework. We treat the interactions of the partons with the medium in the multiple soft scattering approximation. Our main conclusion is that the presence of a mass scale in the problem favors a faster decorrelation of the emitters of the antenna, an effect which leads to an enhancement of the spectrum and is in competition with the known suppression of the spectrum of a single massive emitter. We also compare the cases of a massive quark-gluon and a massive quark-antiquark antennae.

Acknowledgments

This work is supported by European Research Council grant HotLHC ERC-2011-StG-279579, by Xunta de Galicia (Consellería de Educacion); by the Spanish Consolider-Ingenio 2010 Programme CPAN and by FEDER.

References

[1] K. Adcox *et al.* [PHENIX Collaboration], Nucl. Phys. A **757**, 184 (2005); B. B. Back *et al.* [PHOBOS Collaboration], Nucl. Phys. A **757**, 28 (2005); I. Arsene *et al.* [BRAHMS Collaboration], Nucl. Phys. A **757**, 1 (2005); J. Adams *et al.* [STAR Collaboration], Nucl. Phys. A **757**, 102 (2005).
[2] K. Aamodt *et al.* [ALICE Collaboration], Phys. Lett. **B696** (2011) 30-39. [arXiv:1012.1004 [nucl-ex]].
[3] G. Aad *et al.* [Atlas Coll.], Phys. Rev. Lett. **105** (2010) 252303. [arXiv:1011.6182 [hep-ex]].
[4] S. Chatrchyan *et al.* [CMS Coll.], Phys.Rev. **C** 84, (2011) 024906. [arXiv:1102.1957 [nucl-ex]].
[5] R. Baier, Y. L. Dokshitzer, A. H. Mueller, S. Peigne, D. Schiff, Nucl. Phys. **B483** (1997) 291.
[6] R. Baier, Y. L. Dokshitzer, A. H. Mueller, S. Peigne, D. Schiff, Nucl. Phys. **B484** (1997) 265.
[7] B. G. Zakharov, JETP Lett. **63** (1996) 952.
[8] B. G. Zakharov, JETP Lett. **65** (1997) 615.
[9] U. A. Wiedemann, Nucl. Phys. **B588** (2000) 303.
[10] M. Gyulassy, P. Levai and I. Vitev, Nucl. Phys. **B594** (2001) 371.
[11] X. -N. Wang and X. -f. Guo, Nucl. Phys. A **696** (2001) 788 [hep-ph/0102230].
[12] P. B. Arnold, G. D. Moore and L. G. Yaffe, JHEP **0206** (2002) 030 [hep-ph/0204343].
[13] N. Armesto, M. Cacciari, A. Dainese, C. A. Salgado and U. A. Wiedemann. *How sensitive are high-pT electron spectra at RHIC to heavy quark energy loss?*, [arXiv:0511257 [hep-ph]].
[14] L. Apolinario, N. Armesto and C. A. Salgado, Phys. Lett. B **718** (2012) 160 [arXiv:1204.2929 [hep-ph]].
[15] J. -P. Blaizot, F. Dominguez, E. Iancu and Y. Mehtar-Tani, JHEP **1301** (2013) 143 [arXiv:1209.4585 [hep-ph]].
[16] F. D'Eramo, M. Lekaveckas, H. Liu and K. Rajagopal, JHEP **1305** (2013) 031 [arXiv:1211.1922 [hep-ph]].
[17] M. Fickinger, G. Ovanesyan and I. Vitev, JHEP **1307** (2013) 059 [arXiv:1304.3497 [hep-ph]].
[18] A. H. Mueller, Phys. Lett. B **104**, 161 (1981).
[19] B. I. Ermolaev and V. S. Fadin, JETP Lett. **33**, 269 (1981) [Pisma Zh. Eksp. Teor. Fiz. **33**, 285 (1981)].
[20] A. Bassetto, M. Ciafaloni and G. Marchesini, Phys. Rept. **100**, 201 (1983); A. Bassetto, M. Ciafaloni, G. Marchesini and A. H. Mueller, Nucl. Phys. **B207**, 189 (1982).

[21] K. Konishi, A. Ukawa, G. Veneziano, Nucl. Phys. **B157**, 45-107 (1979).

[22] Y. L. Dokshitzer, V. A. Khoze, A. H. Mueller and S. I. Troyan, "Basics Of Perturbative QCD," *Gif-sur-Yvette, France: Ed. Frontieres (1991) 274 p. (Basics of)*

[23] Y. Mehtar-Tani, C. A. Salgado and K. Tywoniuk, Phys. Rev. Lett. **106** (2011) 122002 [arXiv:1009.2965 [hep-ph]].

[24] Y. Mehtar-Tani, C. A. Salgado and K. Tywoniuk, *The radiation pattern of a QCD antenna in a dilute medium*, JHEP **1204** (2012) 064 [arXiv:1112.5031 [hep-ph]].

[25] Y. Mehtar-Tani and K. Tywoniuk, *Jet coherence in QCD media: the antenna radiation spectrum*, JHEP **1301** (2013) 031 [arXiv:1105.1346 [hep-ph]].

[26] N. Armesto, C. A. Salgado and U. A. Wiedemann, *Medium induced gluon radiation off massive quarks fills the dead cone*, Phys. Rev. D **69** (2004) 114003 [hep-ph/0312106].

[27] N. Armesto, H. Ma, Y. Mehtar-Tani, C. A. Salgado and K. Tywoniuk, *Mass effect and coherence in medium-induced QCD radiation off a $q\bar{q}$ antenna*, J. Phys. G **38** (2011) 124140 [arXiv:1107.0291 [hep-ph]].

[28] N. Armesto, H. Ma, Y. Mehtar-Tani, C. A. Salgado and K. Tywoniuk, *Coherence effects and broadening in medium-induced QCD radiation off a massive $q\bar{q}$ antenna*, JHEP **1201** (2012) 109 [arXiv:1110.4343 [hep-ph]].

[29] J. Casalderrey-Solana and E. Iancu, JHEP **1108** (2011) 015 [arXiv:1105.1760 [hep-ph]].

Journal of Physics: Conference Series **589** (2015) 012005 doi:10.1088/1742-6596/589/1/012005

Event characterization in (very) asymmetric collisions

Gabor David

Brookhaven National Laboratory, Upton, NY 11973, USA

E-mail: david@bnl.gov

Abstract. Event-by-event reconstruction of the collision geometry using some incarnation of the Glauber-model is a widely accepted method in studying heavy ion collisions. While there is no known problem with the procedure when applied to the collision of two large ions, we will argue that in very asymmetric collisions, like $p(d)$+A with at least one hard scattering process occuring the event geometry deduced with the simple Glauber-model may be biased.

1. Introduction

Recent results from very asymmetric collisions at RHIC and LHC (so far $p(d)$+A) provided several surprises, questioning the decade-old assumption that such reactions probe the *cold nuclear matter* (CNM) effects only, and as such, serve as reference, quasi-calibration of *no medium* effects, which can then be compared to the A+A collisions, where a hot, dense medium is formed. This prevailing view was, however, fundamentally shaken in 2013. At relatively low transverse momenta, where such phenomena are typically associated with genuine hydrodynamic flow due to a strongly interacting medium, long-range quadrupole azimuthal correlations have been observed at LHC in p+Pb collisions [1, 2, 3] and subsequently in d+Au collisions at RHIC [4]. At the high end of the available p_T range preliminary results [5] indicated a significant change of the nuclear modification factor from peripheral to central d+Au collisions, both for single particles (π^0, η) and reconstructed jets. This result was clearly in tension with earlier findings and theoretical expectations.

In general terms the nuclear modification factor for particle species X and nuclei A, B is defined as

$$R_{BA}^X = \frac{dN_{BA}^X/dp_T dy}{< N_{coll} > dN_{pp}^X/dp_T dy}$$

in case of $p(d)$+A collisions B is simply just one proton or a deuteron. The crucial quantity is N_{coll}, the number of binary nucleon-nucleon collisions in the overlap region of the two species involved. At the very least, N_{coll} depends on the nuclear geometry: the impact parameter of the collision, the density fluctuations of the nucleus, the (energy dependent!) cross section of nucleon-nucleon collisions. In addition other kinematic and dynamic factors may play a role. In this paper we will raise the issue whether the methods to determine collision geometry and subsequent derivation of N_{coll}, worked out and functioning well for collisions of large nuclei, is unquestionably valid for very asymmetric (p/d+A) systems as well, or there are some legitimate concerns? Will the presence of a very hard scattering change the overall event characterization? Can such a change - if it exists - be verified experimentally?

2. Centrality in theory and in the experiment - large A+A collisions

Detailed study of the properties of the sQGP relies heavily on event-by-event classification of the collisions according to the (implied) collision geometry. Theorists need to know the impact parameter b, whose magnitude defines *collision centrality* in the purest sense, in order to calculate the nuclear overlap T_{AB} which then, combined with the nucleon-nucleon cross section lead to quantities like the number of participant nucleons (N_{part}), the number of binary nucleon-nucleon collisions (N_{coll}), as well as the spatial distribution of participating nucleons and quantities derived from it like eccentricity.

Alternately, one can sidestep the question what b is and define the collision centrality with the number of nucleons participating in the collision (N_{part}). Just as b, N_{part} cannot be directly determined in the experiment, it is derived from some simple *global observable* like charged particle multiplicity (N_{ch}) or transverse energy (E_T) in a specific pseudorapidity region. Since we are not discussing any particular experiment, we simply call the detector(s) covering this region and serving both to trigger *minimum bias* events as well as establish collision centrality, Trigger Centrality Detector(s) (TCD). It should be noted that the TCD is usually located close to beam rapidity and far from the rapidity region where the actual signal is measured. The global observables, including the signal in TCD, are correlated with the directly unaccessible N_{part} - the nature of the correlation is discussed extensively in [6] along with a historic overview how our understanding of the underlying processes evolved in the past decades. In case of large colliding nuclei and in an average event (no particles/jets above a few GeV/c observed) the N_{part} vs N_{ch}^{TCD} correlation is quite narrow, so N_{ch}^{TCD} is a reasonable proxy for the unmeasured N_{part} (of b, for that matter). The correspondence is usually established with a Glauber Monte Carlo or with some event-generator [7, 8], by convolving the N_{part} distribution with the (known or assumed) single-collision soft production, and comparing it to the measured N_{ch}^{TCD} distribution. Once a good match between model and experiment is achieved, the measured distribution is divided up to percentiles, and the corresponding $< N_{part} >$, $< N_{coll} >$ established from the model [7].

A crucial fact in large-on-large A+B ion collisions is that the number of participating nucleons[1] from *both* nuclei (N_{part}^A, N_{part}^B) is large in all but the most extremely peripheral collisions. Combined with the observation that soft production per participant pair, the basis to determine centrality, fluctuates, collisions with N_{part}^A, N_{part}^B are practically indistinguishable from collisions with $N_{part}^A, N_{part}^B - 1$. In other words, even if some process would reduce (or outright eliminate) the contribution from one nucleon to N_{ch} production, this would be experimentally undetectable. We believe that this is exactly the reason why the Glauber-model works well on large-on-large systems. The same is not necessarily true when very asymmetric systems collide, like $p(d)$ with a large nucleus A.

3. The case of pp and $p(d) + A$ collisions

It is interesting that in the only experimentally verifiable case - namely pp-collisions - the situation is somewhat more complicated. There are two well-identified issues with N_{ch}^{TCD} production near beam rapidity: the *trigger bias* and the bias on N_{ch}^{TCD} if a high p_T particle is observed far from beam rapidity (say, at $y = 0$). The two effects are demonstrated in a very compact way in Fig. 1 taken from [8]. To trigger an event requires either a coincidence of at least one hit in each of two TCD detectors (up- and downstream), called minimum bias trigger,

[1] We are aware that a recent paper [6] shifts the emphasis from participating nucleons to quark participants, as the scaling variable for N_{ch}, but this doesn't change the essence of our arguments.

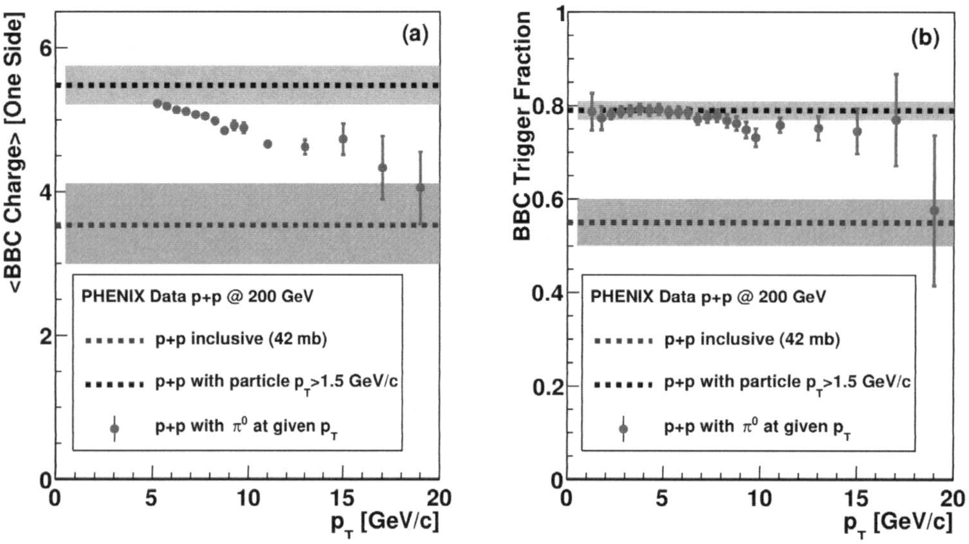

Figure 1. Left (a): N_{ch}^{TCD} at $-3.9 < \eta < -3.1$ vs the highest p_T observed in a single particle at $|\eta| < 0.35$ in pp collisions [8]. The two dashed lines are the mean charge for events taken with minimum bias trigger (lower, blue) and requiring at least one particle with $p_T > 1.5\text{GeV}/c$ at midrapidity. Right (b): Trigger efficiency (probability of the coincidence of at least one particle at both $-3.9 < \eta < -3.1$ and $3.1 < \eta < 3.9$) for minimum bias events (lower, blue line), events with at least one particle with $p_T > 1.5$ at midrapidity (upper, black line), and the dependence on the highest p_T particle observed at midrapidity.

or substantial activity (at least one particle with $p_T > 1.5$) at central rapidity. Note that the later trigger is significantly more efficient, i.e. once there is strong activity at mid-rapidity, *soft* production forward and backward is more likely, too. On the other hand, if the activity at midrapidity is *too* strong (rising maximum p_T), the trigger efficiency drops a few percent, i.e. some of these high p_T events are lost and the loss has to be corrected for.

The left panel (a) is both more dramatic and more relevant for the issue at hand, namely, centrality determination in very asymmetric collisions. It shows the mean value of N_{ch}^{TCD} on one side for minimum bias triggers (blue dashed line), for $p_T > 1.5$ at midrapidity (black dashed line) and as a function of the maximum p_T observed at midrapidity (red points). The exact reason of the relatively fast depletion of N_{ch}^{TCD} with increasing midrapidity p_T is not clear, and herein lies the potential problem in determining centrality in $p(d)$+A collisions from N_{ch}^{TCD}. Obviously in the asymptotic limit - two partons, both carrying $x \approx 1$ fraction of momentum and scattering with the maximum possible q^2 at mid-rapidity - N_{ch}^{TCD} goes to zero, but the probability of this happening is vanishingly small. Still it is useful to keep in mind because at *some* p_T this *kinematic* effect will start playing a role, even if we don't know (in a model-independent, experimentally verifiable way) where. It is, however, unclear whether the kinematic effect is sufficent to explain the entire drop seen in Fig. 1.

Let's turn now to $p(d)$+A collisions. Based on the success of the Glauber Monte Carlo in large A+A collisions it is tempting to apply the same method to derive centrality, and in fact, this is what has been done early on. In case of d+Au collisions there is an added complication from the large size of the deuteron - the two nucleons can be as far apart as 7-8 fm, the radius of the entire Au nucleus - but it also has an advantage: collisions in which only a proton or

Journal of Physics: Conference Series **589** (2015) 012005 doi:10.1088/1742-6596/589/1/012005

neutron interacted can be tagged, providing important cross-checks [8]. There is also a gradual shape-change and shift in rapidity of the N_{ch} distributions with centrality, measured over a wide rapidity range both at RHIC [9] and at LHC [10, 11]. This rapidity shift has been predicted (BGK triangle [12]) and it doesn't prevent us from reliably re-calibrating centrality for the *average* event. To emphasize the word *average* is more than pedantry: a comprehensive review of centrality determination [7] explicitely states that *"In heavy ion collisions, we manipulate the fact that the majority of the initial state nucleon-nucleon collisions will be analogous to minimum bias p+p collisions with a small perturbation from much rarer hard interactions."* In fact, the most authentic source, Prof. Glauber himself cautions the reader in his famous lecture, p.340 in [13]: *"...limitations... the approximate wave function (74) is only adequate for the treatment of small-angle scattering. It does not contain, in general, a correct estimate of the Fourier amplitudes corresponding to large momentum transfer."* This exactly is the justification why in the Glauber model - widely used to establish centrality - all nucleon-nucleon collisions are considered to be independent, occuring with the same cross-section and leading to similar soft particle production, irrespective of their "history"[2].

What happens in *non-average* events, i.e. the ones where at least one very hard scattering occurs? As we've seen in the *pp* case, the TCD multiplicity - the basic building block of determining centrality - changes substantially (see Fig. 1). On the other hand in the usual procedure the measured N_{ch}^{TCD} is compared to an N_{part}-fold convolution of the detector response to an *average* nucleon-nucleon scattering. While this is *not* correct, judging from the experimentally verifiable *pp* case, as long as both N_{part}^{A} and N_{part}^{B} are large, the mistake in assigning a centrality class to the event is minute. However, this is *not* true in $p(d)$+A collisions, since there are only one or two nucleons on one side. Once a hard scattering occured, this nucleon was necessarily part of it. In fact, in [8] the authors dealt with the problem by introducing p_T-dependent centrality bias factors, re-calculating the apparent centrality by folding $N_{coll}-1$-times the *normal* nucleon-nucleon response and once the reduced one. The virtue of this approach is that it changes the generally accepted, commonly used experimental method in just one, clearly defined step. The underlying assumption, while debatable, is crisp: even if *the* projectile nucleon suffered a hard collision, in all other interactions with the target nucleons (calculated from geometry) it will behave as if nothing happened. There are some quantum mechanical (coherence) arguments to justify this assumption, although it is quite remarkable to see them in the context of defending a model whose basic tenet is the incoherent superposition of independent, identical collisions.

Other models take a different route to explain unexpected high-p_T $p(d)$+A results. In an early paper [14] the authors argue that *"Our pA collision study showed that each pp inelastic collision adds ~400 MeV/c transverse momentum to the partons inside the proton (on average). After a few such collisions the partons gain high enough transverse momenta to become free of the proton and during this transition time they do not interact (dead time). We assume that such a proton is "lost" for the reaction and does not participate in particle production anymore. We note that such a picture correponds to a modification of the original Glauber model."*

The color fluctuation model [15] is another attempt to explain the larger than expected N_{coll} fluctuations in p+A collisions. The model, also called Glauber-Gribov model, allows the total nucleon-nucleon cross-section σ_{tot} to fluctuate and thus change the distribution of wounded nucleons at a given impact parameter, or, conversely, the impact parameter distribution for a given number of wounded nucleons. On the experimental side ATLAS [10] used two Glauber-Gribov parametrizations in addition to the standard Glauber to describe their E_T measurement in p+Pb collisions.

[2] I.e. previous collisions and their "violence". We are aware that in a tightly coupled quantum-mechanical system the word "history" may be out of place - but that's exactly how the widely used Glauber model operates.

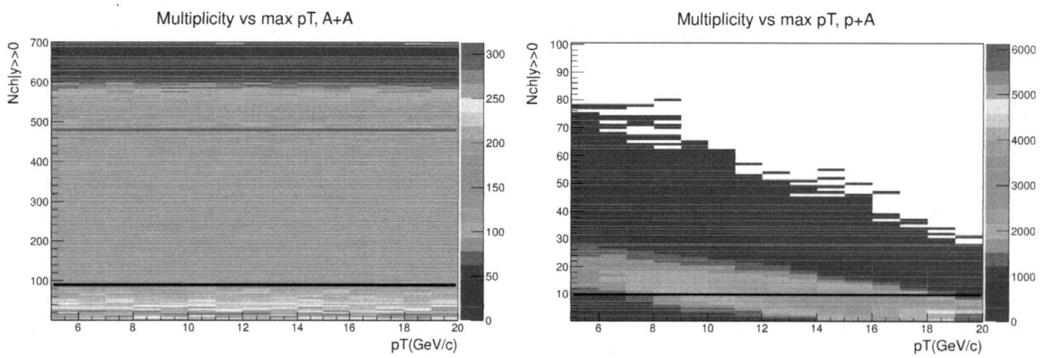

Figure 2. A simple model to illustrate possible differences between large-large and very small - large system collisions. In both panels the vertical axis is a simulated soft multiplicity (N_{ch}^{TCD}) in the region where collision centrality is typically determined. The horizontal axes are the maximum p_T in the event; a hard collision producing high p_T proportionally decreases further soft production from this participating nucleon. Left: large A+A system (here A=50). Right: p+A system, with A=197. Red and black lines: limits for the 20% most central and 20% most peripheral collisions, based on the multiplicity distribution at low maximum p_T (average event).

4. What next?

We believe there is a potentially serious problem in establishing collision geometry in very asymmetric collisions when at least one hard scattering occurred, too. We claim the existence of the problem not because this or that particular measurement, odd or unexpected result (like R_{CP} for identified high p_T particles), but because of the experimental observation in Fig. 1 and the implausibility of the assumption that after a very hard collision the projectile nucleon keeps interacting as an unexcited, unchanged object. For arguments' sake, let's take the opposite extreme and assume that after a very hard scattering - which can turn at any point of the eikonal traversing the target nucleus - the projectile nucleon as a whole is simple "out of the pool", stops interacting. While this is obviously unrealistic (*just as no change whatsoever is*) it's easy to build a toy model around it and check the experimental consequences.

Our toy model is a standard Glauber Monte Carlo, with randomly distributed, fixed size nucleons, the number of participants and collisions are calculated with the hard disk approximation, and the calculated soft production (charged multiplicity) is the N_{part}-fold convolution of a realistic negative binomial distribution (NBD). The only significant difference w.r.t. the standard Glauber model is that in each event we assign one of the collisions of (exactly one) projectile nucleon as hard collision, and the higher the p_T generated, the more we reduce soft production by this nucleon for the rest of its path in the target nucleus. In other words, if in the Glauber-picture the projectile nucleon scattered n times, but the m-th scattering was a hard one producing the maximum p_T in the event, then the total soft production will be calculated as an m-fold convolution of the original NDB and an $n-m$-fold convolution of a reduced NDB, where the mean of the reduced NBD decreases linearly with increasing maximum p_T. Since we are interested in trends, namely, whether there are noticeable changes as the maximum p_T increases, we use a flat distribution for maximum p_T (instead of a realistic spectrum each p_T is thrown with equal probability).

In Fig. 2 we show the result of our toy model for two cases: first, when two large nuclei collide (left panel, A=50), second, when a single proton collides with a large nucleus (right, p+Au). As already pointed out earlier, in large A+B systems, due to the large number of participants in both nuclei (N_{part}^A, N_{part}^B) the fact that one participant from each stops contributing to soft particle production is irrelevant; the fluctuations ensure that the multiplicity distribution - consequently, the multiplicity-based centrality classes - are unchanged irrespective of the maximum p_T observed in the event. The situation is quite different for the p+Au collision (right panel), where with increasing maximum p_T a clear depletion of soft production is observed. We believe such a *triangular* shape with p_T is a general property of very asymmetric systems, while large-on-large systems essentially don't change (they are "rectangular" with p_T). The red and black lines indicate where the 20% most central and 20% most peripheral collisions would be if assigned based on multiplicity in the average (low p_T only) events. While there is no noticeable difference with p_T in A+A events, the character of the events in a centrality bin changes a lot in asymmetric collisions. While this toy model is deliberately simplistic, it describes features seen in actual data quite well.

In summary, we reviewed briefly how fundamental geometry like impact parameter or number of participants in heavy ion collisions are connected to experimental observables. We found that as long as large colliding systems are considered, the correspondence between the theoretical quantities and experimental observables is quite unambigous even in the presence of a few very hard subprocesses. However, in very asymmetric, specifically $p(d)$+A collisions the presence of a hard process strongly biases the soft production, and, as a consequence, the derived geometric quantities as well.

Acknowledgments
The author greatfully acknowledges inspiring discussions with G. Barnafoldi, M. Gyulassy, P. Levai, J. Nagle, N. Novitzky, T. Sakaguchi and M. J. Tannenbaum.

References

[1] Chatrchyan S *et al* 2013 *Phys. Lett.* B **718** 795
[2] Abelev B *et al* 2013 *Phys. Lett.* B **719** 29
[3] Aad G *et al* 2013 *Phys. Rev. Lett.* **110** 182302
[4] Adare A *et al.* 2013 *Phys. Rev. Lett.* **111** 202301
[5] Sahlmueller B, 2013 *Nucl. Phys.* A **904-905** 795c
[6] Adler S S *et al* 2014 *Phys. Rev. C* **89** 044905
[7] Miller M L, Reygers K, Sanders S J, Steinberg P, Annu. Rev. Nucl. Part. Sci **57** (2007) 205
[8] Adare S *et al* 2013 Centrality categorization for $R_{p(d)+A}$ in high-energy collisions *Preprint* nucl-ex/1310.4793
[9] Back B B *et al* 2005 *Phys. Rev.* C **72** 031901
[10] `https://atlas.web.cern.ch/Atlas/GROUPS/PHYSICS/CONFNOTES/ATLAS-CONF-2013-096`
[11] `https://atlas.web.cern.ch/Atlas/GROUPS/PHYSICS/CONFNOTES/ATLAS-CONF-2013-105`
[12] Adil A, Gyulassy M 2005 *Phys. Rev.* C **72** 034907
[13] Glauber R J, in Lectures in Theoretical Physics, ed. Brittin W E, Dunham L G, 1:315. New York: Interscience (1959)
[14] Papp G, Levai P, Barnafoldi G G, Zhang Y, Fai G 2001 *Acta Phys. Polon.* D bf 32 4069 *Preprint* nucl-th/0203075
[15] Alvioli M, Strikman M 2013 *Phys. Lett.* B **722** 347

Journal of Physics: Conference Series **589** (2015) 012006 doi:10.1088/1742-6596/589/1/012006

Modifications of high-p_T di-hadron correlations for identified triggers

Olga Evdokimov *for the STAR Collaboration*
University of Illinois at Chicago, 845 W. Taylor St., Chicago, IL 60607, USA
E-mail: evdolga@uic.edu

Abstract.
Angular correlations with high transverse momentum hadrons have become an established tool for studying properties of the medium created in ultra-relativistic heavy ion collisions. When investigated in two dimensions, relative azimuth and relative pseudorapidity, the small-angle part of such correlations is commonly attributed to QCD jets, while long-range correlations, specifically ridge observed at low- and intermediate momentum, are often described as manifestations of hydrodynamic expansion of the system, e.g. flow. Comparative analysis of both features, presented in this work for correlations with identified leading hadrons, explores the particle type dependence of the jet structures, and challenges the flow-only hypotheses behind the ridge.

1. Introduction

The energy densities achieved in nuclear collisions at the Relativistic Heavy Ion Collider (RHIC) and, more recently, at the Large Hadron Collider (LHC) provided an experimental environment to study new form of QCD matter. The increasing amount of evidence supports the QCD prediction of partonic deconfinement at the limits of high temperature and/or energy density. The creation of the strongly interacting Quark Gluon Plasma (sQGP) is evident, for example, in the collective behavior of the bulk particles in the soft sector (below transverse momentum $p_T < 1.5$ GeV/c). The production rates of identified hadrons (including strange and multi-strange species), the mass-dependent modifications of the spectral distributions, and the strength of azimuthal correlations with respect to reaction plane (elliptic flow, v_2) are all consistently described by a thermalized, hydrodynamically expanding medium with properties resembling a "perfect fluid" [1]. More recently, new theoretical developments led to recognition of the importance of initial state non-uniformity due to density/geometry fluctuations. These fluctuations result in significant anisotropies of higher orders (v_3 and above), while originally it was expected that these terms would be negligible or canceled by symmetry. Investigation of particle-type dependence of v_n anisotropies is one of the goals of this work.

It has been now experimentally established that azimuthal anisotropies extend far above the soft momenta, although different physics mechanisms may be responsible for the effect in different kinematic regimes (see recent review in [2]). The hydrodynamic expansion under pressure gradients dominating the long-range correlations of soft hadrons, gives way to the path-dependent energy loss, or jet quenching, at high p_T. The jet quenching effect in the hard-sector ($p_T > 6$ GeV/c for hadrons, $E_T > 30$ GeV for jets) is also established through the suppression of the hadron and jet production rates at both RHIC and LHC. While little modifications of the surviving jets in the high-p_T regime of heavy ion events compared to the reference measurements from pp collisions has been observed [3, 4], a number of novel phenomena has been discovered at the intermediate momentum range, ($1.5 < p_T < 6$ GeV/c), which is the focus of this study. Of particular interest are prominent excess of the near-side (small relative azimuth, $\Delta\phi$) yields that extends to large relative pseudorapidities ($\Delta\eta$), the ridge. This long-range ridge was discovered in 2-dimensional ($\Delta\eta$, $\Delta\phi$) correlations along with shape modifications of the away-side ($\Delta\phi \sim \pi$) in associated hadron distributions for high-p_T trigger hadrons [5]. A number of physics mechanisms has been initially proposed to describe the ridge, however, recently the most-widely accepted interpretation attributes the phenomenon to the higher order flow terms, the final state imprint of the inhomogeneous initial conditions [6].

Another important discovery of the intermediate momentum range is the observation of constituent quark scaling behavior in the measurements of elliptic flow for baryons and mesons, as well as in the different trends of nuclear modification factors for baryons and mesons [7]. The most natural explanation for the observed trends was put forward by recombination /coalescence models of hadron production [8, 9], in which constituent quark number scaling comes naturally. The recombination approach also provides effortlessly an explanation for the so-called baryon/meson puzzle: it has been observed that hadron composition in central Au+Au collisions at intermediate p_T is very different from pp events, with large relative baryon over meson enhancement across all measured hadron flavors reported.

For the di-hadron correlation measurements in the intermediate momentum range the dominance of hadronization through recombination mechanisms is expected to lead to a "trigger dilution" for the jet-like near-side per-trigger yields measured in Au+Au events compared to the pp reference. That is, the correlations measured in 200 GeV pp collisions for leading hadrons above 4 GeV/c are dominated by the QCD fragmentation contributions and capture the back-to-back di-jet structures left by fragmentation of the hard-scattered partons. In case of recombination of the thermal quarks from the medium created in the Au+Au collisions at the same energy, no near-side jet-like peak would accompany the resulting hadron, thus effectively reducing (diluting) the per-trigger associated yields. Moreover, this dilution effect would be stronger for baryon as compared to meson triggers. There is an earlier measurement of associated near-side yields for protons and pions [10] that shows an increasing difference of the correlation strength for more central events for proton, but not pion triggers, but in this work no separation of the short- and long-range contributions has been performed, complicating the interpretation. For the long-range correlations associated with baryons and mesons, if indeed dominated by the higher order flow terms from hydrodynamic expansion of the fluctuating initial state, one would expect the v_3 (and higher) harmonics to exhibit the same scaling behavior as elliptic flow ($v2$), e.g. stronger correlations for baryons than for mesons. Systematic studies of the two-dimensional di-hadron correlations with various identified leading hadrons are needed to provide further differential tests for establishing model interpretations.

The STAR detector has a great advantage for these types of studies, relying on its full azimuthal and extended pseudorapidity coverage at nearly uniform acceptance. One of the STAR main detectors, the Time Projection Chamber (TPC), provides multiple means to perform particle type identification for the charged hadrons recorded. The momentum information combined with measurements of ionization energy loss in the detector material allows identification of "common" charged hadrons, $\pi^{\pm}$, $K^{\pm}$, and (anti)protons in the soft sector, and, more recently, at higher momenta of the relativistic rise region [11]. Together with invariant mass calculations, this information also allows STAR to topologically identify neutral hadrons decaying weakly into charged particles, particularly, Λ and K^0_s, as well as statistical identification of multiple resonances.

The high purity samples of Λ and K^0_s from the intermediate momentum range have been

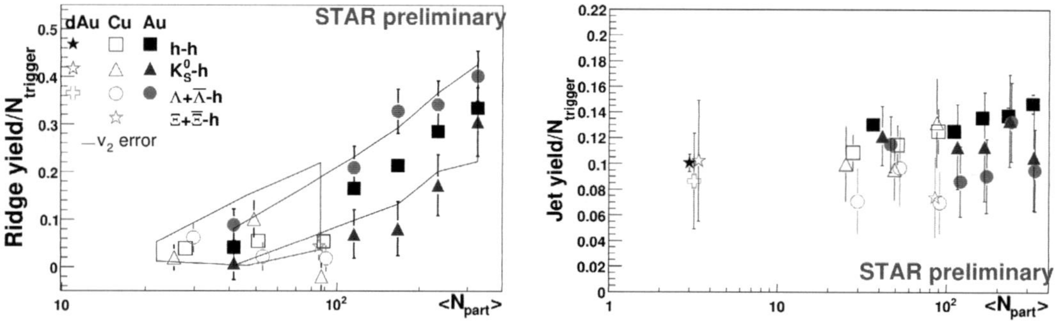

Figure 1. Multiplicity dependence of the ridge-like (left) and jet-like (right) yields associated with different types of identified triggers in 2-dimenssional di-hadron correlations from 200 GeV Au+Au and Cu+Cu collisions of various centralities. All leading hadrons are selected to have transverse momenta $p_T^{\rm trig}$ between 3 and 6 GeV/c, while associated charged hadrons used to build angular correlations are required to have 1.5 GeV/c < $p_T^{\rm assoc}$ < $p_T^{\rm trig}$. The band shows an estimate of systematic uncertainties errors due to subtraction of elliptic flow contributions. The figure is from [12].

used as leading hadrons to construct 2D di-hadron correlations in a first attempt in STAR to quantify the particle-type dependence of correlation strength [12]. In this early work measured correlations have been decomposed into a jet-like part (small $\Delta\phi$, small $\Delta\eta$) and a ridge (small $\Delta\phi$, large $\Delta\eta$) components. The v_2-modulated background have been subtracted from both parts of the correlations for further analysis. The extracted yields have then been studied as a function of event centrality (expressed via number of participants, $N_{\rm part}$, estimated by Glauber model) for Au+Au and Cu+Cu collisions at 200 GeV. It has been reported that while little or no dependence on $N_{\rm part}$ could be seen in the jet-like yields associated with different types of hadronic triggers (Fig. 1, right), the ridge yield increases approximately linearly for all correlations studied (Fig. 1, left). For our discussion the interest is in the comparative analysis of baryon vs. meson related yields. Unfortunately, the uncertainties of these preliminary measurements prevent decisive conclusions. However, we note that the jet-like yields associated with lambdas seem lower than these for kaons, while a reverse tendency is presented in the ridge measurements.

The goal of this work is to revisit the question of particle-type dependence in the correlation structures for leading baryons and mesons taking advantage of the new high luminosity data samples recorded by STAR in the years 2008 and 2010 for 200 GeV d+Au and Au+Au collisions, respectively. We focus on the Au+Au collisions from the top 10% centrality selection, where medium effects are expected to be maximal. The new d+Au dataset is used as a reference establishing a baseline in absence of the hot nuclear matter. Statistical separation of charged pions from (anti)proton and charged kaon triggers, performed in this work for correlation studies, provides new constraints for theoretical description of the data.

2. Analysis of Di-hadron Correlations with Identified Leading Hadrons

The highest p_T charged particle in an event is first selected as a trigger hadron. We require a trigger momentum threshold for all correlations to be between 4 and 5 GeV/c. Statistical hadron identification techniques employing measurements of ionization energy loss in the TPC material in the relativistic rise region are used, following established techniques [11]. The separation between the typical energy deposition (dE/dx) of pions and non-pions in the STAR TPC in the kinematic range of our trigger selection allows trivially obtaining a "pure pion" sample with a

Journal of Physics: Conference Series **589** (2015) 012006 doi:10.1088/1742-6596/589/1/012006

single cut on the dE/dx-related variable. For the pion triggers presented in this work the purity of the sample is estimated to be 98% and contains approximately 50% of all charged pions recorded within the imposed p_T range. The remaining charged hadrons between 4 and 5 GeV/c that fell below our pion selection cut, are then kept as a "pion-depleted" trigger set, which contains majority of (anti)protons and charged kaons, and the remaining 50% of the charged pions. These two trigger samples are used to construct corresponding 2D di-hadron correlations following procedures and corrections established in previous di-hadron works [13]. The charged tracks used to construct the correlations with respect to trigger hadrons are required to have transverse momentum between 1.5 and 4 GeV/c, excluding potential overlap with the leading hadron pool. Figure 2 presents di-hadron correlations, resulting after the efficiency, acceptance, and track splitting/merging effects are taken into account. The left panel shows a baseline correlation made for all leading charged hadrons with 4< p_T^{trig} <5 GeV/c and associated tracks with 1.5< p_T^{assoc} <1 GeV/c without any identification. The middle panel shows correlation obtained with identical kinematic selection for the sample of pure-pion triggers. Even in these raw correlations (without subtraction of combinatorial background) the emerging differences are evident: the jet-like peak appears larger for the pion triggers than that for inclusive charged hadrons, while decrease in the strength of the long range ridge can also be spotted.

Once the 2D correlation for the pion triggers has been measured, this contribution can be directly subtracted from correlation with the pion-depleted sample, which after proper renormalization to account for the number of non-pion triggers left, provides the correlation measurement for proton and kaon mix of leading hadrons ("p+K" or "non-pions" in the following). The relative composition of the non-pion trigger sample is 60% (anti)protons vs. 40% charged kaons. Although further separation of the non-pion sample into pure-protons and pure-kaons is not performed in this work due to significant overlap of the ionization energy loss distributions for these species, comparative analysis of high-precision pure-pion and non-pion triggered correlations allows to infer the differences for baryon vs. mesons, assuming recombination as a dominant mechanism behind constituent quark scaling behavior. No additional treatments or data modeling is needed to note significant differences in both short- and long-range correlation components between the pion and non-pion trigger samples. As seen in the right panel of Fig.2 the non-pion triggered correlations has significantly smaller jet-like peak while the ridge amplitude is much higher than that for pion triggers (as expected from comparison of pion-triggered correlation to that with inclusive charged hadron triggers).

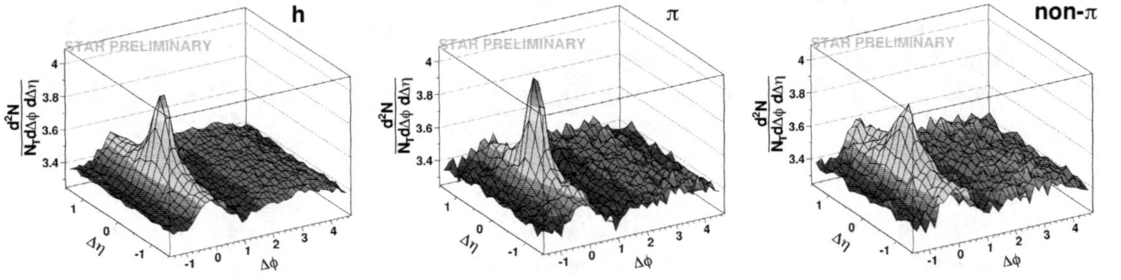

Figure 2. Two-dimensional $\Delta\phi - \Delta\eta$ correlation functions for leading particles with 4< p_T^{trig} <5 GeV/c and associated charged hadrons of 1.5< p_T^{assoc} <4 GeV/c. Unidentified (inclusive) charged hadrons (left), charged pions (middle), and non-pions (right) from 0-10% most central Au+Au collisions at 200 GeV were used as trigger particles for the correlations shown.

3. Results and Discussion

Jet-like peak

We separate the short- and long-range components for each 2D correlation by direct subtraction of the ridge-dominated part (0.9< $|\Delta\eta|$ <1.5) from the small-angle ($|\Delta\eta|$ <0.9) region (after proper normalization per unit $\Delta\eta$). The ridge/jet separation cut at 0.9 ensures that over 98% of jet-like peak is contained within the inner selection. Such subtraction assumes that contributions other than the jet-like peak are $\Delta\eta$-independent; multiple fitting tests performed justify this assumption. The direct subtraction of the ridge region removes the contributions from combinatoric background modulated by hydrodynamic flow, minimizing the uncertainties on the extracted jet-like yields (modulo the assumption of rapidity independence of the flow harmonics in the kinematic region covered). We observe significantly larger jet-like yields associated with pion triggers compared to non-pion measurement. In theoretical calculations the leading pions and non-pions (specifically, protons) have different contributions from fragmentation of quarks and gluons. Thus, it is interesting to test if the color-charge effects in jet-medium interactions can be seen in the correlation data with identified triggers. To differentiate between the cold and hot nuclear matter effects we compare the jet-like yields for pion and non-pion triggers from central Au+Au collisions with the reference measurements obtained in identical way from the d+Au data. We notice no sizable changes in the peak shapes in either $\Delta\eta$ or $\Delta\phi$ dimensions between Au+Au and d+Au data for the same trigger types. The integrated jet-like yield for the pion triggers from central Au+Au data is significantly higher than for pion triggers from d+Au collisions, while non-pion triggers show little to no change. The relative enhancement of the pion-triggered yield is on the order of 30%, and could be attributed to the jet energy loss resulting in additional softer hadrons along the direction of the jet and/or modification of the jet fragmentation pattern due to the medium presence. Qualitatively, this result is consistent with the findings of the jet-track correlation analysis [14]. No change in the jet-like yield for non-pions is unlikely due to color-charge effects, as larger energy loss for leading gluon vs. quark jet would result in the opposite effect, increasing the associated yields even more. The alternative mechanism that would lead to a modification of relative non-pion to pion-triggered yields can be readily provided by the recombination model. The larger contribution from recombination of thermal quarks to proton production compared to that of pions, which expected to be largely formed by through fragmentation by 4 or 5 GeV/c, would result in a trigger dilution effect mentioned earlier. Since the systematic uncertainty on the integrated jet-like yields is dominated by the overall uncertainty on tracking efficiency, which is uncorrelated between the d+Au and Au+Au data samples, but fully correlated for pions and non-pions in the same data, we construct a double ratio of the yields to eliminate this uncertainty source. We find that this double ratio of non-pion to pion triggered jet-like yields from Au+Au data over the same in d+Au is $0.7 \pm 0.1^{\text{stat}}$, pointing to an additional decrease of the associated yields for non-pion triggers in central Au+Au collisions with respect to a reference measurement from d+Au.

Ridge

The dependence of the ridge and away-side yield on trigger identity is another focus of this study. Recently, the ridge and away-side structures are commonly described in terms of higher order Fourier harmonics and rather successfully modeled hydrodynamically. Correlations with identified triggers allow to study scaling features (such as mass or quark number) to support or challenge this picture. To characterize the trigger-type dependence of the long-range components of the measured correlations, the corresponding jet-like near-side peak discussed in the previous section is first subtracted from full 2D measurement. We observe no rapidity-dependent residuals in the obtained correlations, and thus could continue the study with a 1D projection on relative azimuth. Left panel of Fig. 3 overlays the resulting projections for pion and non-pion triggers.

Significantly larger ridge and away-side yields are clearly seen for p+K triggers compared to pion triggers. Fourier fits including harmonics up to fifth order are performed on each of the projections, and are also shown in the Figure. The extracted Fourier coefficients, V_n, each containing the contributions from the corresponding n-th order term for trigger and associated hadrons, are plotted in the middle panel of Fig. 3. For comparison, Fourier coefficients for the correlation in the same kinematic range with unidentified charged triggers are also presented. We find a very small V_5 terms for all correlations, with higher harmonics vanishing completely. For the coefficients V_1 through V_4, the values from non-pion-triggered correlation fit are consistently above those from pion-triggered correlation (with unidentified trigger results placed in between, as expected).

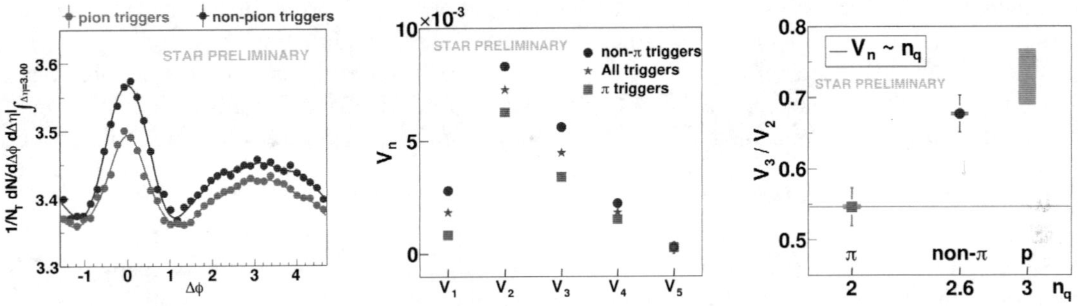

Figure 3. Left: the ridge – $\Delta\phi$ projection of $\Delta\eta$-independent part of the 2D di-hadron correlation for pion and non-pion triggers from 10% most central Au+Au collisions at 200 GeV. The lines illustrate results of the Fourier fit described in the text. Middle: Coefficients of the Fourier fit for long-range correlations associated with pion, non-pion, and inclusive charged trigger particles from the same data. Right: The ratio of the V_3/V_2 coefficients of the Fourier expansion for correlations with pion and non-pion triggers (symbols), and the extrapolated estimate for this ratio for proton triggers (box). The extrapolation assumptions are detailed in the text. The straight line through the pion-related point is intended to guide the eye.

In the hydrodynamic picture the Fourier coefficients extracted from the data fits contain combined information about the flow of the trigger and of associated hadrons. For the separation of these contributions, the v_n factorization is widely accepted, e.g. $V_n =< v_n^{\text{trig}} >< v_n^{assoc} >$. For all correlations studied in this work, the kinematic selection for associated hadrons is identical, therefore one can assume that the associated hadron contributions will cancel in the ratios of the extracted coefficients. Moreover, under the factorization assumption, the constituent quark scaling behavior, if it hold for the third term as well, should lead to the identical ratio of the V_3/V_2 harmonics from each set of the identified triggers, as it would simply reflect per-quark v_3/v_2. The ratios of the V_3/V_2 harmonics extracted from the Fourier decomposition of pion- and non-pion triggered correlations is presented in the right panel of Fig. 3. A significantly larger ratio is observed for p+K triggers, despite consisting of a mix of baryons and mesons, indicating breaking of the scaling for the v_3. For illustration, we extrapolate the measured ratio to baryons-only, assuming (under the constituent quark scaling hypothesis) that the kaon contribution is identical to that of pions, which naturally leads to an even bigger discrepancy. We conclude that triangular harmonic does not follow the same scaling pattern as the elliptic flow. In the context of di-hadron correlation studies, this indicates that the interpretation of the rapidity-independent terms through hydrodynamic flow only is incomplete.

Journal of Physics: Conference Series **589** (2015) 012006 doi:10.1088/1742-6596/589/1/012006

4. Summary

In summary, new measurements of two-dimensional di-hadron correlations with identified leading hadrons from 200 GeV central Au+Au collisions recorded by STAR detector have been presented. For particle identification purposes the relativistic rise of the ionization energy loss in STAR/TPC for charged hadrons between 4 and 5 GeV/c was utilized to separate charged pion triggers. Statistical separation methods were used to extract the correlation measurements for a trigger sample without pions, consisting of about 60% (anti)proton and 40% charged kaon triggers. To study the medium-induced effects on the jet-like part of the correlation, a reference measurement from the minimum bias d+Au events at the same energy has been used. We find that for the pion-led correlation, the associated hadron yield in the jet-like peak is enhanced significantly in central Au+Au collisions compared to d+Au data for the charged hadrons with momenta between 1.5 and 4 GeV/c, studied in this work. At the same time, the non-pion triggers show no such trend for the jet-like peak associated yields. It is hard to expect such differences for the two trigger sets studied to come from color-charge dependence of energy loss, however, the relative decrease of non-pion to pion associated yields in the jet-cone from Au+Au events vs. d+Au events could have a natural explanation from recombination models. For the long-range component of the correlation functions, we observe significantly higher ridge and away-side amplitude associated with the non-pion triggers. Fourier expansion of the long range correlations yields non-zero terms up to the fifth order and allows comparing the constituent quark scaling behavior for the second and higher harmonics. We find that the ratios of V_3/V_2 Fourier coefficients from the fits to the pion-triggered data is significantly lower than from non-pion correlations, indicating the constituent quark scaling pattern is not preserved for the v_3. This new observation strongly suggests that interpretation of the long-range component of the two-dimensional di-hadron correlation in terms of hydrodynamic flow terms only is incomplete.

[1] J. Adams *et al.*, Nucl. Physics **A**757 102 (2005); K. Adcox *et al.*, Nucl. Physics **A**757 184 (2005); B.B. Back *et al.*, Nucl. Physics **A**757 28 (2005); I. Arsene *et al.*, Nucl. Physics **A**757 1 (2005).
[2] S. A. Voloshin, A. M. Poskanzer and R. Snellings, *Collective phenomena in non-central nuclear collisions*, Landolt-Brnstein Database "Relativistic Heavy Ion Physics" Vol. **23** (2010).
[3] J. Adams *et al.*, Phys. Rev. Lett. **97** 162301 (2006); H. Agakishiev *et al.*, Phys. Rev. **C** 83 061901(R) (2011); B. Abelev *et al.*, Phys. Rev. **C** 80 064912 (2009); G. Agakishiev *et al.*, Phys. Rev. **C** 85 014903 (2012).
[4] S. Chatrchyan *et al.*, JHEP **1210** 087 (2012); S. Chatrchyan *et al.*, Phys. Lett. **B** 712 176 (2012).
[5] J. Adams *et al.*, Phys. Rev. Lett. 95 152301 (2005); B. Abelev *et al.*, Phys. Rev. **C** 80 064912 (2009); M. Aggarwal *et al.*, Phys. Rev. **C** 82 024912 (2010). G. Agakishiev *et al.*, Phys. Rev. **C** 85 014903 (2012); S. Chatrchyan *et al.*, Eur. Phys. J. **C** 72 2012 (2012).
[6] B. Alver and G. Roland, Phys. Rev. **C** 81, 054905 (2010).
[7] S. S. Adler *et al.*, Phys. Rev. Lett. **91** 182301 (2003); J. Adams *et al.*, Phys. Rev. Lett. **92** 052302 (2004);
[8] R.C. Hwa and C. B. Yang, Phys. Rev. **C** 67, 064902 (2003); R. J. Fries, B. Muller, C. Nonaka, and S. A. Bass, Phys. Rev. Lett. **90**, 202303 (2003).
[9] V. Greco, C.M. Ko, and P. Levai, Phys. Rev. Lett. **90**, 202302 (2003).
[10] S.S. Adler *et al.*, Phys. Rev. C **71** 051902(R) (2005).
[11] B. I. Abelev *et al.*, Phys. Rev. Lett. **97** 152301 (2006); M. Shao, O. Barannikova, *et al.*, Nucl. Instrum. Meth. **A** 558 419 (2006).
[12] J. Bielcikova, arXiv:0806.2261 [nucl-ex] (2008); G. Agakishiev *et al.*, Phys. Rev. **C** 85 014903 (2012).
[13] M. Aggarwal *et al.*, Phys. Rev. **C** 82 024912 (2010).
[14] L. Adamczyk *et al.*, arXiv:1302.6184 [nucl-ex] (2013).

Journal of Physics: Conference Series **589** (2015) 012007 doi:10.1088/1742-6596/589/1/012007

The Initial Flow of Classical Gluon Fields in Heavy Ion Collisions

Rainer J Fries and Guangyao Chen
Cyclotron Institute and Department of Physics and Astronomy, Texas A&M University, College Station TX 77843, USA
E-mail: rjfries@comp.tamu.edu

Abstract. Using analytic solutions of the Yang-Mills equations we calculate the initial flow of energy of the classical gluon field created in collisions of large nuclei at high energies. We find radial and elliptic flow which follows gradients in the initial energy density, similar to a simple hydrodynamic behavior. In addition we find a rapidity-odd transverse flow field which implies the presence of angular momentum and should lead to directed flow in final particle spectra. We trace those energy flow terms to transverse fields from the non-abelian generalization of Gauss' Law and Ampére's and Faraday's Laws.

1. Introduction

At asymptotically large energies (or small Bjorken-x) the gluon fields in hadrons and nuclei approach a state of nuclear matter generally referred to as color glass condensate (CGC) [1, 2]. The transverse gluon density saturates at a value $\sim Q_s^{-2}$ characterized by a saturation scale Q_s. In this limit large occupation numbers permit a quasi-classical treatment of the gluon field which is the approximation known as the McLerran-Venugopalan (MV) model [3, 4]. The saturation scale Q_s grows with the size of the nucleus $\sim A^{1/3}$. Hence high energy nuclear collisions at the Relativistic Heavy Ion Collider (RHIC) and the Large Hadron Collider (LHC) offer unique opportunities to study CGC.

Here we report on results from a calculation which solves the classical gluon field (Yang-Mills) equations employing an expansion in powers of the proper time $\tau = \sqrt{t^2 - z^2}$ after the collision. We focus on the transverse Poynting vector $S^i = T^{0i}$, $i = 1, 2$ of the gluon field, where $T^{\mu\nu}$ is the energy momentum tensor. S^i describes the initial transverse flow of energy of the gluon field, but we expect this flow to translate into hydrodynamic flow of quark gluon plasma once the system thermalizes. Our results have first been reported in detail in [5, 6].

2. Color Glass Condensate

We seek solutions of the classical Yang-Mills equations

$$[D_\mu, F^{\mu\nu}] = J^\nu \tag{1}$$

where the $SU(3)$-current J^μ is generated by two transverse charge distributions $\rho_1(\vec{x}_\perp)$ and $\rho_2(\vec{x}_\perp)$ moving on the + and − light cone respectively. The ρ_i describe the $SU(3)$ charge distributions of the large-x partons in the colliding nuclei which generate the small-x gluon field. Typically these equations are solved numerically in the forward light cone ($\tau \geq 0$) [7, 8, 9].

However, it is possible to obtain an analytic recursive solution as well, as first described in [10]. To that end one employs a power series for the gauge potential A^μ in the forward light cone,

$$A(\tau, \vec{x}_\perp) = \sum_{n=0}^{\infty} \tau^n A_{(n)}(\vec{x}_\perp)\,, \tag{2}$$

$$A^i_\perp(\tau, \vec{x}_\perp) = \sum_{n=0}^{\infty} \tau^n A^i_{\perp(n)}(\vec{x}_\perp)\,, \tag{3}$$

where in light cone notation $A^\pm = \pm x^\pm A$. The leading terms are given by boundary conditions on the light cone [11]

$$A^i_{\perp(0)} = A^i_1 + A^i_2\,, \tag{4}$$

$$A_{(0)} = -\frac{ig}{2}\left[A^i_1, A^i_2\right]\,, \tag{5}$$

where A^i_1 and A^i_2 are the gauge potentials in nucleus 1 and 2 (generated by charges ρ_1 and ρ_2), respectively, before the collision. The recursion relation for coefficients of even powers n ($n > 1$), are

$$A_{(n)} = \frac{1}{n(n+2)} \sum_{k+l+m=n-2} \left[D^i_{(k)}, \left[D^i_{(l)}, A_{(m)}\right]\right]\,, \tag{6}$$

$$A^i_{\perp(n)} = \frac{1}{n^2}\left(\sum_{k+l=n-2} \left[D^j_{(k)}, F^{ji}_{(l)}\right] + ig \sum_{k+l+m=n-4} \left[A_{(k)}, \left[D^i_{(l)}, A_{(m)}\right]\right]\right)\,, \tag{7}$$

while all odd coefficients vanish. The convergence radius of this series in τ will generally be of order $1/Q_s$, but for example the series recovers the known solution for the abelian case for all times [12].

The physical fields $F^{\mu\nu}$ can be expanded in τ as well and can be computed order by order from A^μ. The dominant fields for small times, i.e. at order $\mathcal{O}(\tau^0)$, are the longitudinal chromo-electric and -magnetic fields [10, 13]

$$E_{(0)} = F^{+-}_{(0)} = ig\left[A^i_1, A^i_2\right]\,, \tag{8}$$

$$B_{(0)} = F^{21}_{(0)} = ig\epsilon^{ij}\left[A^i_1, A^j_2\right]\,. \tag{9}$$

From this result one can calculate an initial energy density $\epsilon_0 = T^{00}_{(0)}$. After averaging over charge densities ρ_i one obtains an event-averaged energy density [14, 12]

$$\varepsilon_0(\vec{x}_\perp) = 2\pi\alpha_s^3 \frac{N_c}{N_c^2 - 1}\mu_1(\vec{x}_\perp)\mu_2(\vec{x}_\perp)\ln^2\frac{Q^2}{\hat{m}^2}\,. \tag{10}$$

Here the usual assumptions of the MV model have been applied, i.e. the ρ_i follow Gaussian distributions and μ_1 and μ_2 determine the average squared charge distribution in each nucleus. Q and $\hat{m}$ are UV and IR cutoffs respectively.

Journal of Physics: Conference Series **589** (2015) 012007 doi:10.1088/1742-6596/589/1/012007

3. Transverse Fields and Transverse Flow

Transverse electric and magnetic fields enter at order $\mathcal{O}(\tau^1)$ in the power series of $F^{\mu\nu}$. The corresponding coefficients can be computed to be [6]

$$E^i_{(1)} = -\frac{1}{2}\left(\sinh\eta[D^i, E_0] + \cosh\eta\, \epsilon^{ij}[D^j, B_0]\right) \tag{11}$$

$$B^i_{(1)} = \frac{1}{2}\left(\cosh\eta\, \epsilon^{ij}[D^j, E_0] - \sinh\eta[D^i, B_0]\right) . \tag{12}$$

Here η is the space-time rapidity. One can readily verify that these expressions are simply the $SU(3)$ analogs of Gauss' Law and Faraday's and Ampére's Laws. More specifically, the η-odd terms emerge naturally as a consequence of Gauss' Law, see Fig. 1.

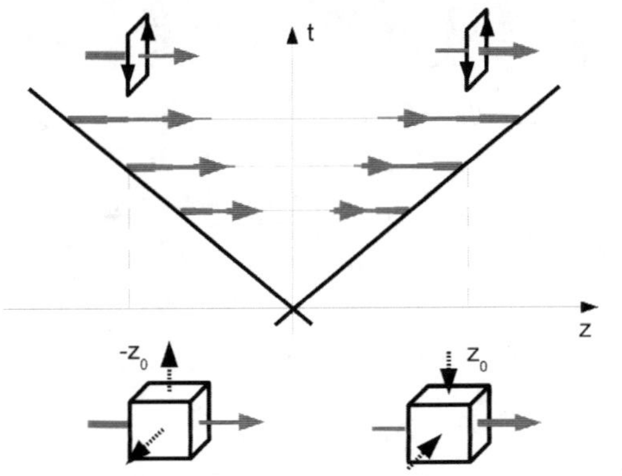

Figure 1. Schematic sketch of Gauss' Law (cubes) and Ampére's and Faraday's Laws (loops) for positive and negative values of the longitudinal coordinate z. The initial longitudinal electric and magnetic fields in the forward light cone diminish as functions of t and z. Transverse fields due to Gauss' Law will have different signs for positive and negative z, while transverse fields due to Ampére's and Faraday's Law are symmetric in z. The abelian version is shown here, see [6] for details.

Fig. 2 shows a typical configuration of transverse fields (in the abelian case for simplicity) for randomly seeded longitudinal fields E_0 and B_0. At rapidity $\eta = 0$ the transverse fields are divergence-free, i.e. only closed field lines due to Ampére's and Faraday's Law appear. At $\eta = 1$ contributions from Gauss' Law are present as well.

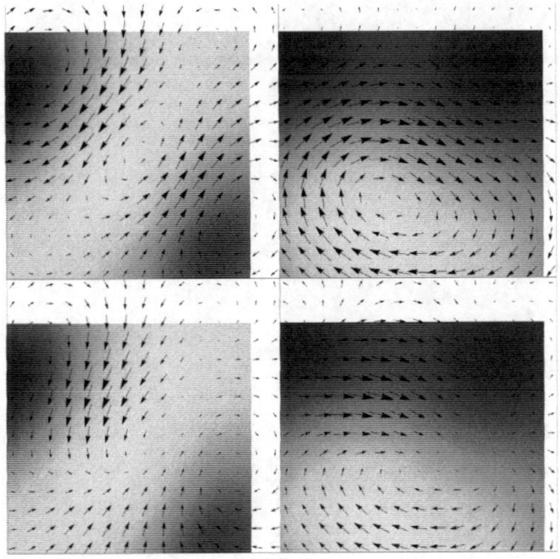

Figure 2. Transverse electric fields (left panels) and magnetic fields (right panels) at $\eta = 0$ (upper panels) and $\eta = 1$ (lower panels) for a random distribution of initial longitudinal electric (shown as background in the right panels) and magnetic fields (left panels). At $\eta = 0$ the fields are divergence-free. The abelian case is shown here, see [6] for details.

We are now able to calculate the flow of energy due to transverse fields. The transverse Poynting vector S^i receives its first contribution in the power series of $T^{\mu\nu}$ at order $\mathcal{O}(\tau^1)$. I.e.

transverse flow in color glass sets in linearly in time,

$$T^{0i} = \frac{\tau}{2} \cosh\eta\, \alpha^i + \frac{\tau}{2} \sinh\eta\, \beta^i . \qquad (13)$$

There is a rapidity-even part α^i and a rapidity-odd contribution β^i to the Poynting vector which are given by [6]

$$\alpha^i = -\nabla^i \varepsilon_0 \quad , \qquad \beta^i = \epsilon^{ij} \left([D^j, B_0] E_0 - [D^j, E_0] B_0 \right) . \qquad (14)$$

The rapidity-odd flow is expected from the existence of rapidity-odd gauge fields, however it is more of a surprise when one approaches the topic of early flow from a purely phenomenological point of view. α^i corresponds to energy flow as expected from hydrodynamic expansion, following the gradient of energy. On the other hand β^i is determined by the gauge field structure underlying the energy density.

After averaging over color charges ρ_i we obtain event-averaged transverse flow fields [6]

$$\alpha^i = -\varepsilon_0 \frac{\nabla^i (\mu_1 \mu_2)}{\mu_1 \mu_2} \quad , \qquad \beta^i = -\varepsilon_0 \frac{\mu_2 \nabla^i \mu_1 - \mu_1 \nabla^i \mu_2}{\mu_1 \mu_2} . \qquad (15)$$

We omit the notation $\langle \ldots \rangle$ for event-averaged quantities if no confusion can arise. Fig. 3 shows the initial event-averaged transverse flow field in Au+Au collisions at impact parameter $b = 6$ fm for two space-time rapidities η.

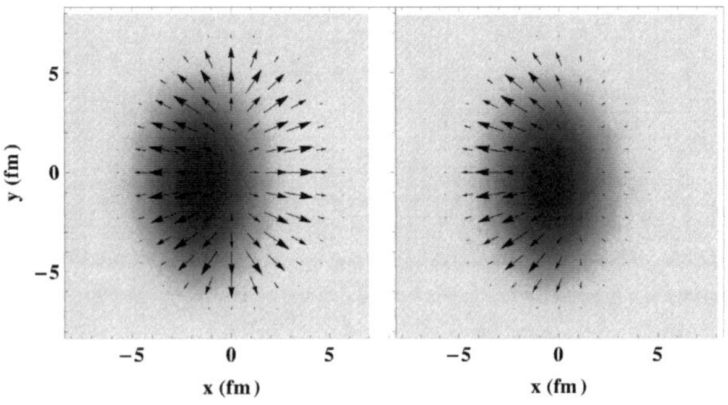

Figure 3. Transverse flow of energy T^{0i} (black arrows; arbitrary units) and energy density ε_0 (shading) in the transverse plane for Au+Au collisions at impact parameter $b = 6$ fm. The nucleus with center located at $x = 3$ fm travels into the plane (the positive η-direction). Left Panel: $\eta = 0$. Right Panel: $\eta = 1$.

One can clearly see a radial and elliptic flow pattern for $\eta = 0$ which is qualitatively similar to what would develop in hydrodynamics. At $\eta = 1$ the rapidity-odd flow β^i becomes comparable to α^i and we notice that the gluon energy flow develops a preferred direction. This is akin to directed flow v_1. We can see this more clearly in Fig. 4 where the transverse flow is shown as a function of x and η in the plane $y = 0$. The gluon field expands more rapidly in the wake of a passing nucleus.

Interesting flow patterns can also be observed for asymmetric collision systems. Fig. 5 shows the transverse flow in the $x - \eta$-plane for Au+Cu collisions for two different impact parameters. Once again energy flow is larger in the wake of the spectators of the larger nucleus, leading to a strong asymmetry between forward and backward rapidities.

4. Summary and Discussion

We have calculated the initial flow of energy of the classical gluon field in high energy nuclear collisions. Our results are valid up to roughly a time $\sim 1/Q_s$ after the collision. The gluon field is expected to decohere and thermalize soon after. Energy and momentum conservation will

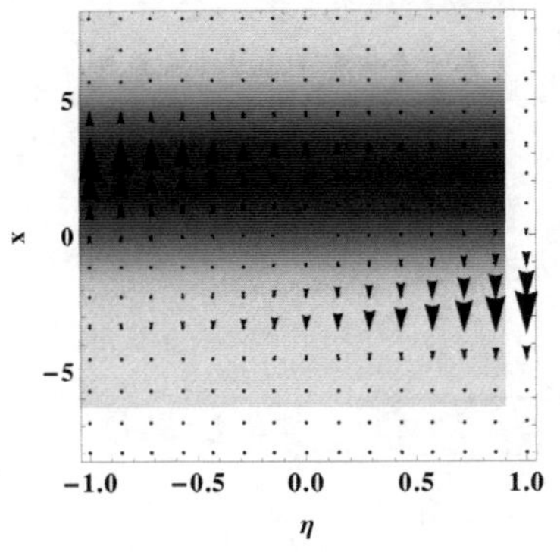

Figure 4. Same as Fig. 3 but plotted in the $\eta - x$-plane defined by $y = 0$. A flow pattern akin to directed flow emerges. The nucleus with center at $x = 3$ fm is traveling to the right.

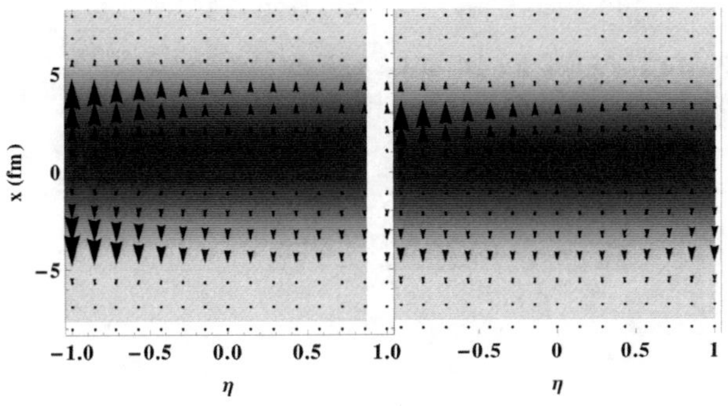

Figure 5. The same as Fig. 4 for Au+Cu (Au traveling to the right). Left Panel: $b = 0$ fm. Right Panel: $b = 2$ fm.

translate the flow fields found here into hydrodynamic flow which then will develop further and eventually freeze out into particle flow. We expect the rapidity-odd pre-equilibrium flow β^i to translate into (rapidity-odd) directed flow v_1 of particles. The sign and general rapidity dependence of β^i are qualitatively consistent with experimental results [6] but a more quantitative statement would require a follow-up 3+1-D viscous hydrodynamic simulation.

One can also interpret the flow term β^i for symmetric collision systems with finite impact parameters b with the inevitable presence of angular momentum L_y (perpendicular to the reaction plane). The initial gluon field transfers a part of the angular momentum in the system before the collision onto the fireball after the collision [15, 16]. There it leads to a rotation of the fireball and possibly to vorticity in the quark gluon fluid.

Another very interesting result is the flow found for asymmetric A+B collision systems. The field β^i specifically relies on the fact that classical gauge fields are the relevant degrees of freedom and one could speculate that observables exist in asymmetric collision systems which are unique signatures for the flow of gauge fields. A further investigation into this direction would be worthwhile.

In the future we plan to use our results on pre-equilibrium flow in viscous hydrodynamic calculations. Preliminary results show that key features of $T^{\mu\nu}$, like angular momentum, readily translate into hydrodynamic fields (local energy density,fluid velocity, shear stress, etc.) in a rapid thermalization scenario [17, 9].

Acknowledgments
This work was supported by the U.S. National Science Foundation through CAREER grant PHY-0847538, and by the JET Collaboration and DOE grant DE-FG02-10ER41682.

Journal of Physics: Conference Series **589** (2015) 012007 doi:10.1088/1742-6596/589/1/012007

References

[1] Iancu E and Venugopalan R 2004 *Quark gluon plasma* vol 3, eds. R C Hwa and X. N. Wang (World Scientific) p 249 (*Preprint* hep-ph/0303204)
[2] Gelis F, Iancu E, Jalilian-Marian J and Venugopalan R 2010 *Ann. Rev. Nucl. Part. Sci.* **60** 463
[3] McLerran L D and Venugopalan R 1994 *Phys. Rev.* D **49** 3352
[4] McLerran L D and Venugopalan R 1994 *Phys. Rev.* D **49** 2233
[5] Chen G and Fries R J 2013 *J. Phys.: Conf. Ser.* **446** 012021
[6] Chen G and Fries R J 2013 *Phys. Lett.* B **723** 417
[7] Krasnitz A and Venugopalan R 2001 *Phys. Rev. Lett.* **86** 1717
[8] Lappi T 2003 *Phys. Rev.* C **67** 054903
[9] Schenke B, Tribedy P and Venugopalan R 2012 *Phys. Rev. Lett.* **108** 252301
[10] Fries R J, Kapusta J I and Li Y 2006 *Preprint nucl-th/0604054*
[11] Kovner A, McLerran L D and Weigert H 1995 *Phys. Rev.* D **52** 3809
[12] Chen G, Fries R J, Kapusta J I, Li Y 2014 *In preparation*
[13] Lappi T and McLerran L 2006 *Nucl. Phys.* A **772** 200
[14] Lappi T 2006 *Phys. Lett.* B **643** 11
[15] Liang Z T and Wang X N 2005 *Phys. Rev. Lett.* **94** 102301, *Erratum-ibid.* **96** 039901
[16] Csernai L P, Magas V K, Stocker H and Strottman D D 2011 *Phys. Rev.* C **84** 024914
[17] Fries R J, Kapusta J I and Li Y 2006 *Nucl. Phys.* A **774** 861

Journal of Physics: Conference Series **589** (2015) 012008 doi:10.1088/1742-6596/589/1/012008

Hard probes and the event generator EPOS

B Guiot and K Werner

SUBATECH, University of Nantes-IN2P3/CNRS-EMN, Nantes, France

E-mail: guiot@subatech.in2p3.fr

Abstract. After a short presentation of the event generator EPOS, we discuss the production of heavy quarks and prompt photons which has been recently implemented. Whereas we have satisfying results for the charm, work on photons is still in progress.

1. Introduction

Today, there is a large amount of data from p-p, Pb-Pb and recently p-Pb collisions at the LHC which need to be interpreted. EPOS, based on 'Parton-based Gribov-Regge theory' [1], aims to reproduce a large range of LHC observables like jets, multiplicity or collective behavior. We will discuss the recent implementation of charm and prompt photons in this event generator. We want hard probes production to be under control for p-p collisions and then, use them for the study of the QGP.

First, we will quickly show the general features of EPOS. Then the charm production will be detailed and finally our projet on prompt photons will be exposed.

2. General presentation

Some important features of EPOS are :

(i) Being a real event generator
(ii) Multiple interactions based on a quantum formalism
(iii) Perturbative calculation with resummation of collinear corrections at the order $\left(\alpha_s(Q^2)\ln(\frac{Q^2}{\mu^2})\right)^n$
(iv) Core-corona separation
(v) Hydrodynamics done event by event
(vi) Hadronisation done using a string fragmentation model for the core

By "being a real event generator", we mean that one event in the LHC $\simeq$ one event in EPOS. The program will generate pions even if one is only interested in charm. All particles are registered in a table and, at the end, one has to select particles of interest.

Our model for multiple interactions is based on a marriage of Gribov-Regge theory [2, 3] and pQCD. It gives the possibility of a quantum treatment of multiple interactions. By "based on Gribov-Regge theory" we mean an assumption about the structure of the T matrix, expressed in

Journal of Physics: Conference Series **589** (2015) 012008 doi:10.1088/1742-6596/589/1/012008

terms of elementary objects called Pomerons (not the same object in EPOS and Gribov-Regge theory). The total cross section can be expressed as illustrated on figure 1.

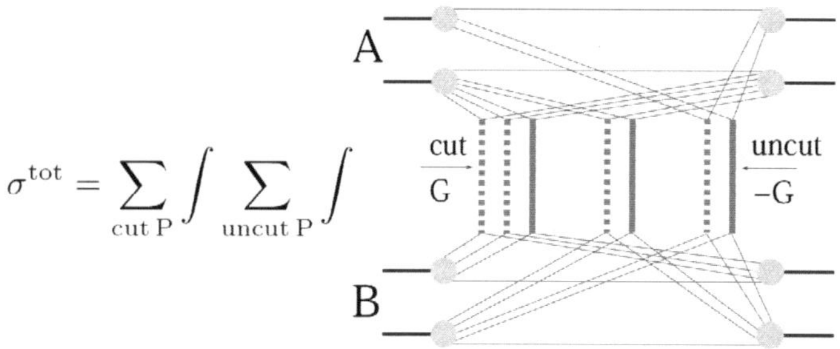

Figure 1. Pink lines : pomerons. A and B are nuclei. Small horizontal lines are remnants.

Partial summation provides exclusive cross sections. In Gribov-Regge theory, the elastic amplitude is given by :

$$A_{2\rightarrow 2}(s,t) = \sum_n A_n(s,t) \tag{1}$$

with $A_n(s,t)$ corresponding to the amplitude for n pomeron(s) exchange. Following the same idea, we can defined σ_m corresponding to the cross section for m cut pomerons. A cut pomeron, figure 6, is at the origin of particles production. The multiplicity is then, on the average, proportional to :

$$N \propto m \sum_m \sigma_m \tag{2}$$

The treatment is the same for p-p, p-A or A-A collisions.

The hydrodynamical evolution is done event by event. Initial conditions are given by the distribution of cut pomerons which correspond to color flux tubes, figure 2. Flux tubes fragment into string pieces which will later constitute particles. These flux tubes will constitute both bulk matter (if the energy density is high enough) and jets. "Matter" is defined by the region of high energy density flux tubes (the blue region figure 3). Then, there are 3 possibilities :

(i) The string piece (the red one) is formed outside the "matter". In that case, it simply escapes as a jet.

(ii) The string piece (the pink one) is formed inside the "matter" but has not enough energy to escape. It constitutes the "matter" and will evolve with the hydrodynamical code.

(iii) The string piece (the blue one) is formed inside the "matter" and has enough energy to escape (based on energy loss argument). It escapes as a jet which has interacted with the fluid.

For more details, see [4, 5].

With these prescriptions, we can reproduce the v_2 for identified particles or the ridge, even for p-Pb collisions (figure 4 and figure 5).

Journal of Physics: Conference Series **589** (2015) 012008 doi:10.1088/1742-6596/589/1/012008

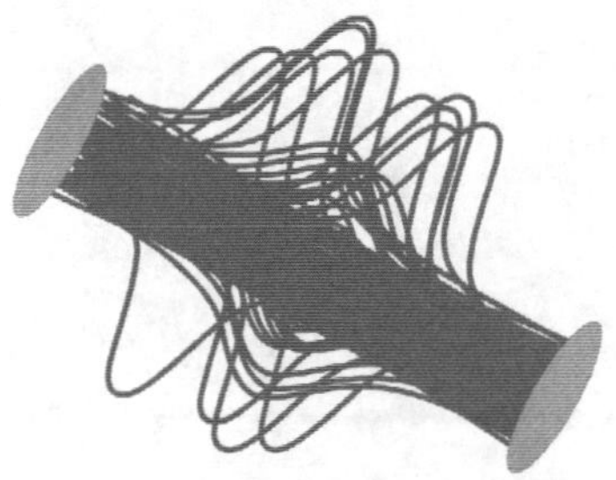

Figure 2. Cut pomerons form color flux tubes between the 2 nuclei.

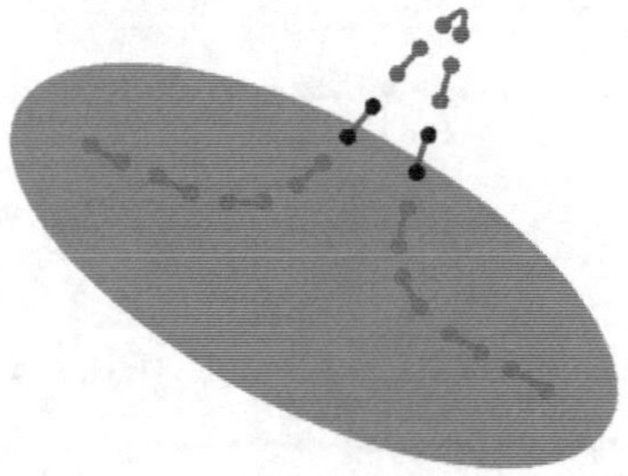

Figure 3. Color flux tubes fragment into string piece. The blue region is "matter" formed by high energy density flux tubes.

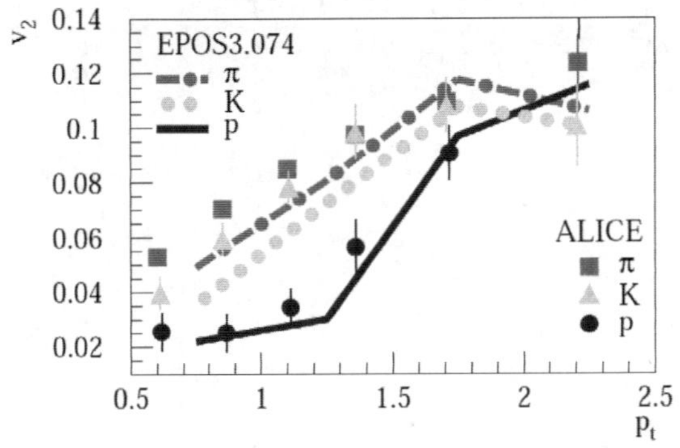

Figure 4. (Color online) Elliptical flow coefficients v2 for pi- ons, kaons, and protons. We show ALICE results (squares) and EPOS3 simulations (lines). Pions appear red, kaons green, protons blue.

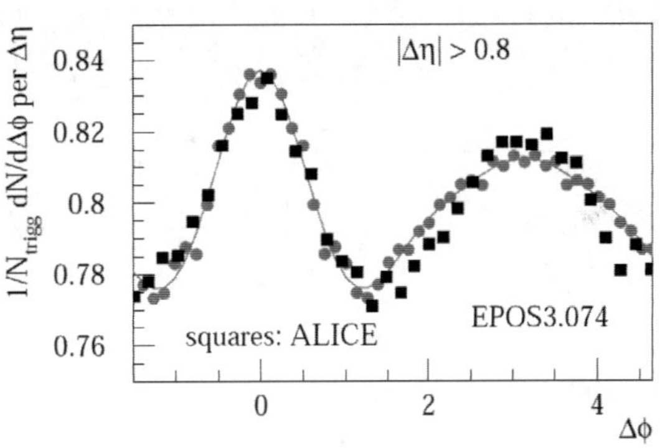

Figure 5. (Color online) Associated yield per trigger, pro- jected onto $\Delta\phi$, for $|\Delta\eta| > 0.8$. We show ALICE results (black squares) and EPOS3 simulations (red dots).

3. Charm production

A charm can be produced during the spacelike cascade, the born process, the timelike cascade (partonic shower) and the string fragmentation, see figures 6 and 7. Based on DGLAP formalism, spacelike and timelike cascades resumme collinear divergences.

During the spacelike cascade, a spacelike parton emits timelike particles until he reaches the born process. In the leading log approximation, virtualities are strongly ordered :

$$Q_1^2 \ll Q_2^2 \ll ... \ll Q_{born}^2 \tag{3}$$

We use the following probability distribution for variables Q^2 and x :

$$\frac{dP(Q_0^2, Q^2, x)}{dxdQ^2} \propto \frac{\alpha_s}{2\pi}\frac{p(x)}{Q^2}\Delta(Q_0^2, Q^2) \qquad Q_0^2 < Q^2 \tag{4}$$

where $p(x)$ are the appropriate splitting functions and x is the fraction of the parent parton light cone momentum k^+ :

$$k^{'+} = xk^+ \tag{5}$$

Journal of Physics: Conference Series **589** (2015) 012008 doi:10.1088/1742-6596/589/1/012008

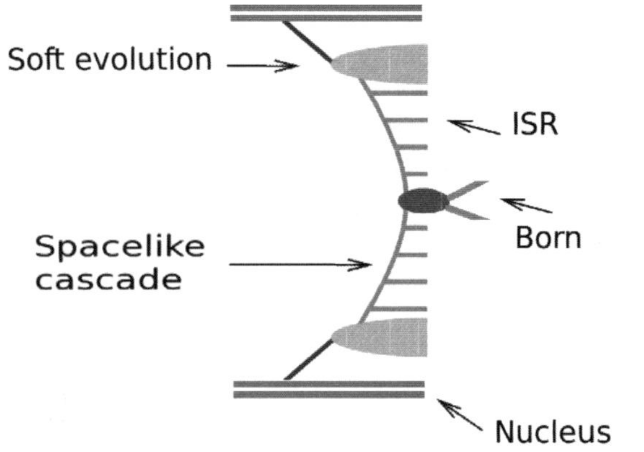

Figure 6. A cut pomeron. The center of the pomeron is a (pQCD) ladder diagram.

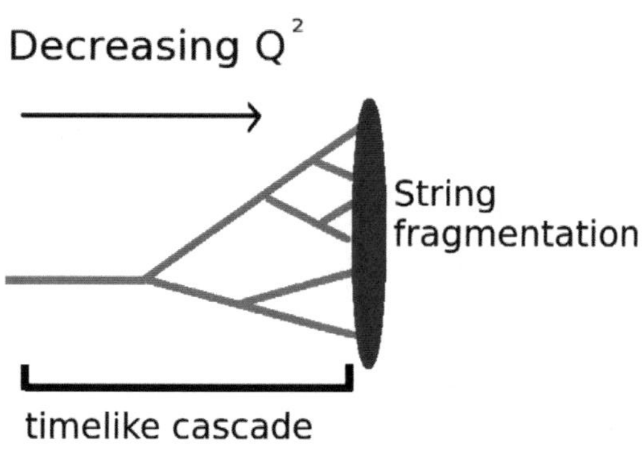

Figure 7. Partonic shower and hadronization by string fragmentation.

$\Delta(Q_0^2, Q^2)$, the Sudakov form factor, gives the probability for no resolvable emissions between Q_0^2 and Q^2. In text books one can find :

$$r^2 = \frac{(1-x)}{x}Q^2 - \frac{p_t^2}{x} \tag{6}$$

r being the 4-momentum of the emitted (timelike) parton and p_t its transverse momentum. Usually one takes $r^2 = 0$, but for heavy quarks we choose :

$$r^2 = m^2 \tag{7}$$

which implies :

$$x < \frac{Q^2}{Q^2 + m^2} \tag{8}$$

The phase space for radiating a heavy quark is smaller than the one for light partons.

The born process, in the center of the ladder, is nothing else that the leading order cross sections for α_s^2, $\alpha_s\alpha_{el}$ and α_{el}^2 processes.

Outgoing on-shell partons ($r^2 = 0$) is an approximation. Out-born partons and those emitted during the spacelike cascade have a finite virtuality :

$$Q^2_{max} \sim p_t^2 + m^2 \tag{9}$$

During the timelike cascade (partonic shower) these partons will loose their virtuality by doing successive splittings. This cascade is stopped when $Q^2 \sim \Lambda^2_{QCD}$. The leading log approximation is used and angular ordering [6, 7] is implemented. The emission probability is :

$$\frac{dP(Q_0^2, Q^2, z)}{dxdQ^2} \propto \frac{\alpha_s}{2\pi}\frac{p(z)}{Q^2}\Delta(Q_0^2, Q^2) \qquad Q_0^2 > Q^2 \tag{10}$$

z being the splitting variable defined as :

$$z = E_{children}/E_{parent} \tag{11}$$

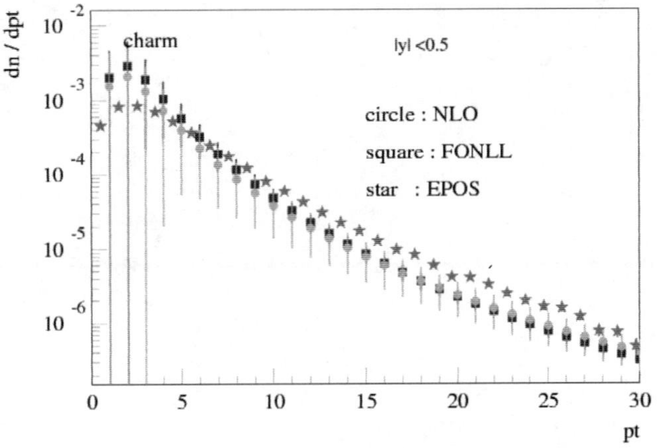

Figure 8. charm from Cacciari vs EPOS.

For the implementation of charm production, no parameters have been changed. Our first test is the comparison with FONLL calculations of M. Cacciari [8], figure 8. At high p_t the shape is the same but our central values are higher. Next, we test EPOS for $D+$ and $D0$ mesons, by comparing our results with the Alice experiment [9] and FONLL calculation, figure 9 and figure 10.

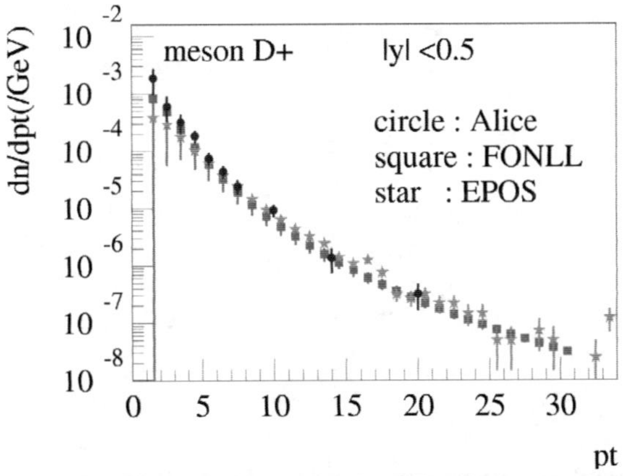

Figure 9. D+ yield from EPOS compare to FONLL calculation [8] and Alice experiment [9].

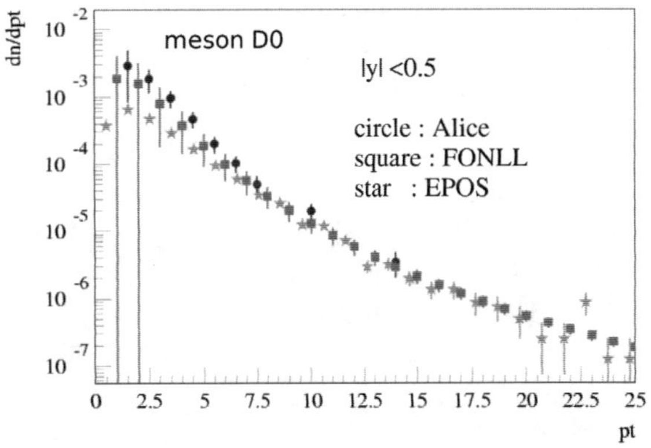

Figure 10. D0 yield from EPOS compare to FONLL calculation [8] and Alice experiment [9].

EPOS uncertainties account only for statistics and there is no charm production during the string fragmentation. Our results are globally in good agreement with Alice and FONLL. However, charm quarks are missing at very low p_t. The explanation could be that charm production during the timelike cascade is too small. In the game of fitting data, we can reproduce Alice results by allowing charm production in string fragmentation (one parameter is changed), figure 11.

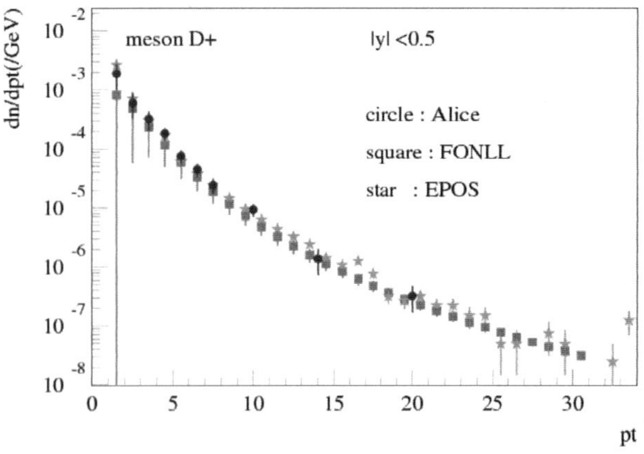

Figure 11. D+ from EPOS with $c\bar{c}$ creation during strings fragmentation.

4. Prompt photons

We want to :

(i) Study isolation criteria

(ii) Compare EPOS and Jetphox

(iii) Use EPOS for γ/jet and $\gamma/hadron$ correlations

Photons production in the born process was already implemented, but it was not the case for the spacelike and the timelike cascade. It has been done with the same formalism used for partons i.e based on the probability eq. 4. One needs to replace α_s by α_{el} and use the appropriate splitting function. The splitting of a photon into a pair of particle-antiparticle is neglected.

Like in experiments, we have an isolation condition on selected photons. Using the table where final particles are registered, we define a cone of radius R around the photon candidate. If the sum of transverse energy of particles in this cone is smaller than a given value (5 GeV for CMS [10]), then the photon is isolated. Here, the fact that EPOS is experiment like is important. In Jetphox, the addition of this isolation criteria gives a non-physical rise of the cross section with $1/R$ [11].

Work on photons is still in progress. Fragmentation photons are strongly suppressed due to isolation requirement, whereas $\sim 98\%$ of direct photons are isolated, table 1.

Comparison with CMS [10], figure 12, shows that our yield seems to be too low by approximately a factor of 1.5. Comparison with Jetphox will give precious information which, I hope, will allow us to improve our results for photons.

Table 1. Pourcentage of isolated direct photons as a function of transverse mometum for the isolation criteria used by CMS [10] ($R = 0.4$, $\sum p_t < 5$GeV).

p_t	# of isolated direct photons/# of direct photons
11	0.972
13	0.981
15	0.971
17	0.973
19	0.976
21	0.992
23	0.947
25	0.991
27	0.976
29	0.966
31	0.977
33	0.978
35	0.978

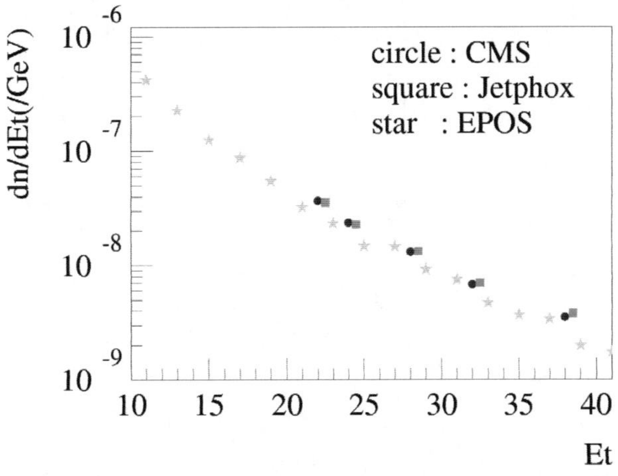

Figure 12. Isolated photons. For more clarity, a +0.5 shift (x axis) is used for Jetphox data.

5. Conclusion

A unified formalism in EPOS is one of its strengths. Heavy quarks and prompt photons production are based on the same equations with the same parameters. Whereas the work on charm is nearly finished, prompt photons physics still need to be improved. We already have very good results for flow [5, 12], and with the implementation of hard probes, EPOS could be an excellent tool for the study of the QGP.

Acknowledgement

I would like to thank the ***projet TOGETHER, region Pays de la Loire*** which has financed my thesis.

References

[1] Drescher H J, Hladik M, Ostapchenko S, Pierog T and Werner K 2001 *Phys. Rep.* **350** 93
[2] Gribov V N 1968 *Sov. Phys. JETP* **26** 414
[3] Werner K 1993 *Phys. Rep.* **232** 87

Journal of Physics: Conference Series **589** (2015) 012008 doi:10.1088/1742-6596/589/1/012008

[4] Werner K, Karpenko I, Bleicher M, Pierog T and Porteboeuf-Houssais S 2012 *arXiv 1203.5704v2*
[5] Werner K, Bleicher M, Guiot B, Karpenko I and Pierog T 2013 *arXiv 1307.4379v1*
[6] Webber B R 1984 *Nuclear Physics B* **238** 492
[7] Marchesini G and Webber B R 1984 *Nuclear Physics B* **238** 1
[8] Cacciari M, Greco M and Nason P 1998 *arXiv hep-ph/9803400*
[9] Alice collaboration 2012 *arXiv 1111.1553v2*
[10] CMS collaboration 2011 *arXiv 1012.0799v2*
[11] Catani S, Fontannaz M, Guillet J Ph and Pilon E 2002 *arXiv hep-ph/0204023*
[12] Werner K, Guiot B, Karpenko I and Pierog T 2013 *arXiv 1312.1233*

Journal of Physics: Conference Series **589** (2015) 012009 doi:10.1088/1742-6596/589/1/012009

Azimuthal elliptic anisotropy (v_2) of high-p_T direct γ in Au+Au collisions at $\sqrt{s_{NN}} = 200$ GeV

Ahmed M. Hamed (STAR Collaboration)
University of Mississippi, Oxford, USA,
Texas A&M University, College Station, USA
E-mail: amhamed@olemiss.edu, ahamed@comp.tamu.edu

Abstract. Preliminary results from the STAR collaboration for the azimuthal elliptic anisotropy (v_2) of high transverse momentum (p_T) direct photons (γ_{dir}) produced at mid-rapidity ($|\eta^{\gamma_{dir}}| < 1$) in Au+Au collisions at center-of-mass energy $\sqrt{s_{NN}} = 200$ GeV are presented, and compared to the measured (v_2) of neutral pions (π^0) in the same kinematic range. The electromagnetic transverse shower profile is used to distinguish π^0 from direct photons. The measured $v_2^{\gamma_{dir}}(p_T)$ at high p_T ($8 < p_T^{\gamma_{dir}} < 20$ GeV/c) is found to be smaller than that of π^0 and consistent with zero when using the forward detectors in determining the event plane.

1. Introduction

The azimuthal distribution of the produced particles in heavy-ion collisions is expected to be sensitive to the initial geometric overlap of the colliding nuclei, and would result in anisotropic azimuthal distributions with respect to the event plane. The standard method to quantify the azimuthal elliptic anisotropy is to expand the particle azimuthal distributions in a Fourier series $\frac{dN}{d\phi}(p_T) = \frac{N}{2\pi}[1 + \sum_n 2v_n(p_T)\cos(n(\phi_{p_T} - \psi_{\rm EP}))]$, where ϕ_{p_T} is the azimuthal angle of the produced particle with certain value of p_T, $\psi_{\rm EP}$ is the azimuthal angle of the event plane, and v_n is the coefficient of the nth harmonic. The 2nd Fourier moment ($n = 2$) is referred to as the "elliptic flow" parameter in the context of the hydrodynamical descriptions, v_2 and its differential form is given by

$$v_2(p_T) = \langle\langle e^{2i(\phi_{p_T} - \psi_{\rm EP})}\rangle\rangle = \langle\langle \cos 2(\phi_{p_T} - \psi_{\rm EP})\rangle\rangle, \tag{1}$$

where the brackets denote statistical averaging over particles and events.

While RHIC data show large amount of elliptic flow as predicted by the hydrodynamic models at low p_T, the results at high p_T [1] are not expected to follow hydrodynamic behavior. The medium-induced radiative energy loss of partons (jet-quenching) has been proposed as the source for the large observed azimuthal elliptic anisotropy at high p_T, due to the path-length dependence of the parton energy loss [2]. The STAR results [3] show the amount of v_2 at high p_T is larger than the predicted values by pure jet-quenching models. Although recent measurements by PHENIX [4] show the produced π^0's in-plane outnumber those produced out-of-plane which may be consistent with the path-length dependence of energy loss, the event plane determination might have remaining bias toward the direction of the produced jets. On the other hand, STAR

Journal of Physics: Conference Series **589** (2015) 012009 doi:10.1088/1742-6596/589/1/012009

results on the suppression of direct γ-triggered vs. π^0-triggered correlated yields [5] show no sensitivity to the path length dependence of parton energy loss. If the $v_2^{\gamma_{dir}}$ can be measured without bias in the event-plane determination, then the measured value can help disentangle the various scenarios of direct photon production through the expected opposite contributions to the v_2 [6, 7, 8, 9], and therefore could help to confirm the observed binary scaling of the direct photon [10].

2. Analysis and Results

2.1. Electromagnetic neutral clusters

The STAR detector is well suited for measuring azimuthal angular correlations due to the large coverage in pseudorapidity and full coverage in azimuth (ϕ). While the Barrel Electromagnetic Calorimeter (BEMC) [11] measures the electromagnetic energy with high resolution, the Barrel Shower Maximum Detector (BSMD) provides fine spatial resolution and enhances the rejection power for the hadrons. The Time Projection Chamber (TPC: $|\eta| < 1$) [12] identifies charged particles, measures their momenta, and allows for a charged-particle veto cut with the BEMC matching. The Forward Time Projection Chamber (FTPC: $2.4 < |\eta| < 4.0$) [13] is used to measure the charged particles momenta and to reconstruct the event plane angle in this analysis. Using the BEMC to select events (*i.e.* "trigger") with high-p_T γ, the STAR experiment collected an integrated luminosity of 535 μb^{-1} of Au+Au collisions in 2007 and 973 μb^{-1} of Au+Au collisions in 2011. In this analysis, events having primary vertex within ± 55 cm of the center of TPC along the beamline, and with at least one electromagnetic cluster with $E_T > 8$ GeV are selected. More than 97% of these clusters have deposited energy greater than 0.5 GeV in each layer of the BSMD. A trigger tower is rejected if it has a track with $p > 3.0$ GeV/c pointing to it, which reduces the number of the electromagnetic clusters by only $\sim 7\%$.

2.2. v_2 of neutral and charged particles

The v_2 is determined using the standard method (Eq. 1), which correlates each particle with the event plane determined from all charged particles with $p_T < 2$ GeV/c (minus the particle of interest). The event plane is determined by

$$\psi_{\rm EP} = \frac{1}{2}\tan^{-1}\left(\frac{\sum_i \sin(2\phi_i)}{\sum_i \cos(2\phi_i)}\right), \tag{2}$$

where ϕ_i are the azimuthal angles of all the particles used to define the event plane. In this analysis, the charged-track quality criteria are similar to those used in previous STAR analyses [14]. The event plane is measured using different techniques and detectors: 1) using all the selected tracks inside the TPC (full-TPC), 2) using the selected tracks in the opposite pseudorapidity side to the particle of interest (off-η), and 3) using all tracks inside the FTPC (full-FTPC) in order to reduce the "non-flow" contributions (azimuthal correlations not related to the event plane). Since the event plane is only an approximation to the true reaction plane, the observed correlation is divided by the event plane resolution. The event plane resolution is estimated using the sub-event method in which the full event is divided up randomly into two sub-events (for full-TPC, off-η, and full-FTPC separately) as described in [15]. Biases due to the finite acceptance of the detector, which cause the particles to be azimuthally anisotropic in the laboratory system are removed according to the method in [16].

2.3. Transverse shower profile analysis

A crucial part of the analysis is to discriminate between showers from γ_{dir} and two close γ's from high-p_T π^0 symmetric decays. At $p_T^{\pi^0} \sim 8$ GeV/c, the angular separation between the two γ's resulting from a π^0 decay is small, but a π^0 shower is generally broader than a single γ

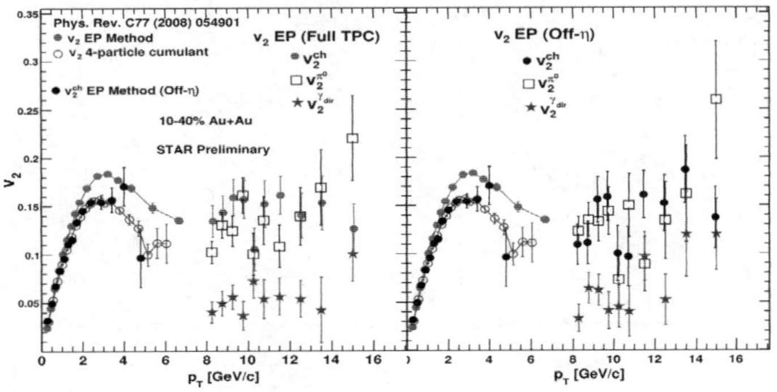

Figure 1. (Color online) For $p_T < 6$ GeV/c, (Au+Au 2007) both panels show previous STAR measurements [14] of v_2 as a function of p_T for charged particles with $|\eta| < 1$ in 10-40% $Au + Au$ collisions at $\sqrt{s_{NN}} = 200$ GeV using the Event-Plane method (closed red circles), and the 4-particle cumulant method (open circles). Also v_2 for charged particles ($|\eta| < 1$) using off-η event plane method is shown in closed black circles (this analysis). For $p_T > 6$ GeV/c: (Au+Au 2007) both panels show v_2 of charged particles, π^0, and γ_{dir} (circles, squares, stars respectively) using the full TPC (left panel) and using the off-η method (right panel).

shower. The BSMD is capable of $(2\gamma)/(1\gamma)$ separation up to $p_T^{\pi^0} \sim 20$ GeV/c due to its high granularity ($\Delta\eta \sim 0.007$, $\Delta\phi \sim 0.007$). The shower shape is quantified as the cluster energy, measured by the BEMC, normalized by the position-weighted energy moment, measured by the BSMD strips [5]. The shower profile cuts were tuned to obtain a nearly γ_{dir}-free (π^0_{rich}) sample and a sample rich in γ_{dir} (γ_{rich}). Since the shower-shape analysis is only effective for rejecting two close γ showers, the γ_{rich} sample contains a mixture of direct photons and contamination from fragmentation photons (γ_{frag}) and photons from asymmetric hadron (π^0 and η) decays.

2.4. v_2 of direct photons

The $v_2^{\gamma_{dir}}$ is given by:

$$v_2^{\gamma_{dir}} = \frac{v_2^{\gamma_{rich}} - \mathcal{R} v_2^{\pi^0}}{1 - \mathcal{R}}, \tag{3}$$

where $\mathcal{R}=\frac{N^{\pi^0}}{N^{\gamma_{rich}}}$, and the numbers of π^0 and γ_{rich} triggers are represented by N^{π^0} and $N^{\gamma_{rich}}$ respectively. The value of $\mathcal{R}$ is measured in [5] and found to be $\sim 30\%$ in central Au+Au. In Eq. 3 all background sources for γ_{dir} are assumed to have the same v_2 as π^0. Thus, all remaining background is subtracted under the assumption that the background particles have the same correlation functions as that measured for π^0 triggers. Figure 1 (both panels) shows the v_2 of charged particles (v_2^{ch}) at low p_T ($p_T < 6$ GeV/c) using the event plane method (off-η) compared to previous STAR measurements [14], and the v_2 of the charged particles, neutral pions and direct photons using the full-TPC and off-η event plane methods at high p_T ($p_T > 6$ GeV/c). At low p_T the v_2^{ch}(off-η) is smaller than the v_2 using the full TPC and agrees well with the $v_2\{4\}$ (4-particle cumulant) method, in which the contribution of the non-flow is expected to be small. At high p_T the two different methods (full TPC and off-η) for the event plane measurements give similar results , which might indicate that the off-η method is not free from a bias in the event-plane determination. While the $v_2^{\pi^0}$ and v_2^{ch} are similar ($\sim 12\%$), the $v_2^{\gamma_{dir}}$ is systematically lower than that of hadrons. The similarity of the v_2 results using the

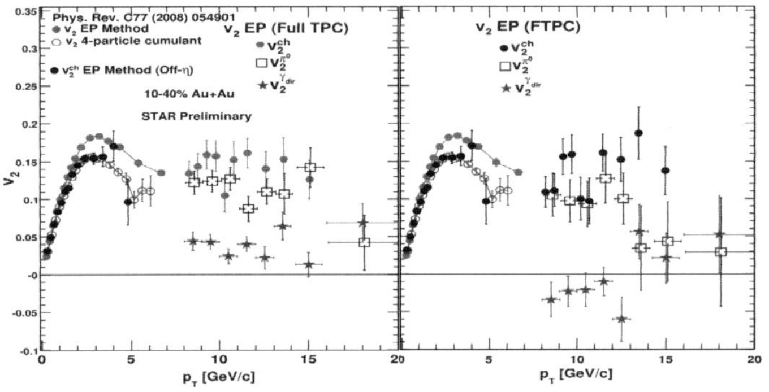

Figure 2. (Color online) For $p_T < 6$ GeV/c, (Au+Au 2007) both panels show measurements as in Fig. 1. For $p_T > 6$ GeV/c, (Au+Au 2011 for π^0 and γ_{dir}) both panels show v_2 of charged particles, π^0, and γ_{dir} (circles, squares, stars respectively) using the full TPC (left panel) and using full FTPC (right panel).

full-TPC and off-η at high p_T, along with the non-zero value of $v_2^{\gamma_{dir}}$, indicate a remaining bias in the event-plane determination.

Figure 2 (left and right panels) shows the $v_2^{\pi^0}$ and $v_2^{\gamma_{dir}}$ for ($8 < p_T^{\gamma_{dir}} < 20$ GeV/c) from Au+Au 2011 data using the full-TPC ($|\eta| < 1$) and full-FTPC ($2.4 < |\eta| < 4.0$). The results from two different data sets (Au+Au 2007 and Au+Au 2011) are consistent (left panels of Fig. 1 and Fig. 2) using the full-TPC. While using the FTPC in determining the event plane (Fig. 2 - right panel) the $v_2^{\gamma_{dir}}$ is consistent with zero. Assuming the dominant source of direct photons is prompt hard production, the zero value implies no remaining bias in the event-plane determination. Accordingly, the measured value of $v_2^{\pi^0}$ would be the effect of path-length dependent energy loss. Systematic studies are currently in progress.

3. Conclusions

The STAR experiment has reported the first $v_2^{\gamma_{dir}}$ at high-p_T ($8 < p_T^{\gamma_{dir}} < 20$ GeV/c) at RHIC. Using the mid-rapidity detectors in determining the event plane, the measured value of $v_2^{\gamma_{dir}}$ is non-zero, and is probably due to biases in the event-plane determination. Using the forward detectors in determining the event plane could eliminate remaining biases, and the measured $v_2^{\gamma_{dir}}$ is consistent with zero. The zero value of $v_2^{\gamma_{dir}}$ suggests a negligible contribution of jet-medium photons [7], and negligible effects of γ_{frag} [6] on the $v_2^{\gamma_{dir}}$ over the covered kinematics range. The measured value of $v_2^{\pi^0}$, using the forward detectors in determining the event plane, is apparently due to the path-length dependence of energy loss.

References

[1] STAR Collaboration, J. Adams *et al* , Nucl. Phys. A**757**, 102 (2005).
[2] E. V. Shuryak, Phys. Rev. C **66** 027902 (2002).
[3] STAR Collaboration, J. Adams *et al* , Phys. Rev. Lett. **93**, 252301 (2004).
[4] PHENIX Collaboration, A. Adare *et al* , Phys. Rev. Lett. **105**, 142301 (2010).
[5] STAR Collaboration, B. I. Abelev *et al* , Phys. Rev. C **82**, 034909 (2010).
[6] B. G. Zakharov, JETP Lett. **80**, 1 (2004).
[7] R. J. Fries, B. Müller, and D. K. Srivastava, Phys. Rev. Lett. **90**, 132301 (2003).
[8] S. Turbide, C. Gale, and R. J. Fries, Phys. Rev. Lett. **96**, 032303 (2006).
[9] G. Qin *et al* , Phys. Rev. C **80** 054909 (2009).
[10] PHENIX Collaboration, S. S. Adler *et al* , Phys. Rev. Lett. **94**, 232301 (2005).

[11] M. Beddo *et al* , Nucl. Instrum. Meth. A**499**, 725 (2003).
[12] M. Anderson *et al* , Nucl. Instrum. Meth. A**499**, 659 (2003).
[13] K. H. Ackermann *et al* , Nucl. Instrum. Meth. A **499**, 713 (2003).
[14] STAR Collaboration, B. I. Abelev *et al* , Phys. Rev. C **77** 054901 (2008).
[15] A. M. Poskanzer and S. A. Voloshin , Phys. Rev. C 58, **1671** (1998).
[16] E877 Collaboration, J. Barrette *et al* , Phys. Rev. C **56** 3254 (1997).

Journal of Physics: Conference Series **589** (2015) 012010 doi:10.1088/1742-6596/589/1/012010

Centrality and rapidity dependence of inclusive pion and prompt photon production in p+Pb collisions at the LHC with EPS09s nPDFs

I. Helenius[1,2], K. J. Eskola[1,2] and H. Paukkunen[1,2]

[1]Department of Physics, P.O. Box 35 (YFL), FI-40014 University of Jyväskylä, Finland
[2]Helsinki Institute of Physics, P.O. Box 64, FI-00014 University of Helsinki, Finland

E-mail: ilkka.helenius@jyu.fi

Abstract. The centrality dependencies of the inclusive neutral pion and prompt photon nuclear modification factors for p+Pb collisions at the LHC are studied using a spatially dependent set of nuclear PDFs, EPS09s. The calculations are performed at mid- and forward rapidities searching for an observable which would optimally probe the spatial dependence of the nuclear PDFs. In addition, we discuss to which x values of the nucleus the different observables are sensitive.

1. Introduction

In the framework of collinear factorization [1, 2], the cross section for producing a hard parton k in a nuclear collision can be calculated as

$$\mathrm{d}\sigma^{AB\to k+X} = \sum_{i,j,X'} f_i^A(x_1,Q^2)\otimes f_j^B(x_2,Q^2)\otimes \mathrm{d}\hat{\sigma}^{ij\to k+X'} + \mathcal{O}(1/Q^2), \tag{1}$$

where $f_i^A(x,Q^2)$s are the nuclear parton distribution functions (nPDFs) describing the number distributions of parton i at given values of momentum fraction x and scale Q for nucleus A. The partonic component $\mathrm{d}\hat{\sigma}^{ij\to k+X'}$ can be calculated using perturbative QCD (pQCD). The key assumption here is that, in the same way as the free proton PDFs, the non-perturbative nPDFs are process independent, so that they can be determined from a certain collection of experimental data and later on used to study different processes. It has been observed that the nPDFs, if they exist, are modified with respect to the free proton PDFs [3, 4, 5, 6]. This modification can be made explicit by decomposing the nPDFs as

$$f_i^A(x,Q^2) = R_i^A(x,Q^2)\, f_i^N(x,Q^2), \tag{2}$$

where $f_i^N(x,Q^2)$ is the free nucleon PDF and $R_i^A(x,Q^2)$ the corresponding nuclear modification. Introducing a non-perturbative input at the initial scale Q_0 and using the DGLAP evolution equations one can then determine $R_i^A(x,Q^2)$ via standard procedures of global analyses. Nowadays, there are several nPDF sets on the market which are reviewed e.g. in Refs. [7, 8].

Each collision of two heavy nuclei takes place at a specific impact parameter $\mathbf{b}$, which is the vector in the transverse plane between the centers of the colliding nuclei. The value of the impact

Journal of Physics: Conference Series **589** (2015) 012010 doi:10.1088/1742-6596/589/1/012010

parameter determines the geometry (centrality) of the collision: if the impact parameter is small the amount of interacting nuclear matter is large (central collision) and if it is large, only the edges of the nuclei collide (peripheral collision). The inclusive hard particle yield in a nuclear collision is typically assumed to be proportional to the amount of interacting matter which is given by the nuclear overlap function $T_{AB}(\mathbf{b})$:

$$T_{AB}(\mathbf{b}) = \int \mathrm{d}^2\mathbf{s}\, T_A(\mathbf{s}+\mathbf{b}/2)\, T_B(\mathbf{s}-\mathbf{b}/2), \tag{3}$$

where $T_A(\mathbf{s})$ is the nuclear thickness, obtained by integrating the nuclear density profile over the beam direction.

As the thickness of the nucleus is not constant in the transverse plane, one would presume that also the nPDFs should somehow depend on the position of the nucleon inside the nucleus. However, all global fits of nPDFs have considered only minimum bias collisions (no cuts on centrality) and hence these cannot be fully consistently used for more restricted centrality classes. In Ref. [9] we addressed this problem by developing a model framework for spatially dependent nPDFs which is described in the next section. Then, in section 3, we present some predictions concerning the centrality dependence for inclusive neutral pion and prompt photon production in p+Pb collisions at the LHC at different rapidities, discuss which observable would be ideal to constrain the spatial dependence of the nPDFs, and which values of x we probe at different rapidities. More predictions for p+Pb collisions at the LHC with other approaches can be found from compilation [10].

2. Framework

A natural requirement for the spatially dependent nuclear modifications $r_i^A(x, Q^2, \mathbf{s})$ is that they should reduce to the minimum bias modification $R_i^A(x, Q^2)$ upon taking the spatial average,

$$R_i^A(x, Q^2) \equiv \frac{1}{A} \int \mathrm{d}^2\mathbf{s}\, T_A(\mathbf{s})\, r_i^A(x, Q^2, \mathbf{s}). \tag{4}$$

This restriction does not, however, constrain the functional form of the spatial dependence in any way, and we need to make an assumption for this. We have chosen a power series of the nuclear thickness function:

$$r_i^A(x, Q^2, \mathbf{s}) = 1 + \sum_{j=1}^{n} c_j^i(x, Q^2)\, [T_A(\mathbf{s})]^j. \tag{5}$$

In this ansatz, all the A and $\mathbf{s}$ dependencies are dictated by the thickness function and the fit parameters $c_j^i(x, Q^2)$ depend only on x and Q^2. Thus, we can obtain the values for $c_j^i(x, Q^2)$ by using an existing set of minimum bias nuclear modifications through minimizing χ^2 defined as

$$\chi_i^2(x, Q^2) = \sum_A \left[\frac{R_i^A(x, Q^2) - \frac{1}{A}\int \mathrm{d}^2\mathbf{s}\, T_A(\mathbf{s})\, r_i^A(x, Q^2, \mathbf{s})}{W_i^A(x, Q^2)} \right]^2, \tag{6}$$

where the sum runs over several different nuclei A. We have performed this procedure for the EKS98 [11] and EPS09 [3] globally analyzed nPDF sets and we refer to the fits obtained here as EKS98s and EPS09s, respectively. For EKS98s we used the relative error as a weight ($W_i^A(x, Q^2) = 1 - R_i^A(x, Q^2)$) and for EPS09s this was set to unity. For both of these sets we were able to obtain accurate fits with four first non-trivial terms in the power series ($n = 4$) in the whole kinematic region considered, see an example in Figure 1. The fitting was performed also for the EPS09 error sets to assess the uncertainties. An example of the resulting spatially dependent nuclear modification is plotted in Figure 2 below.

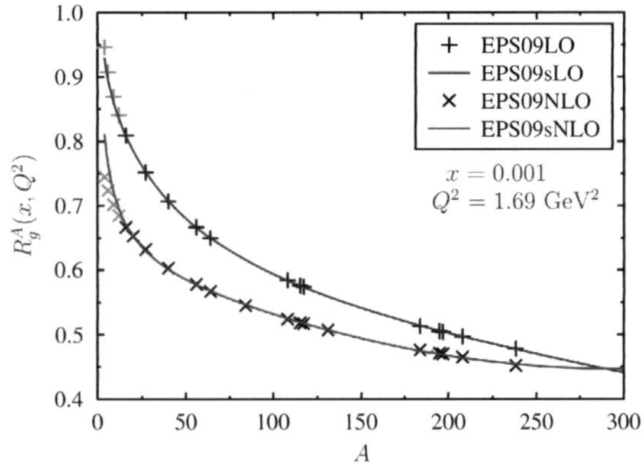

Figure 1. The spatially averaged (EPS09s) and the minimum bias (EPS09) nuclear modifications for gluons with fixed x and Q^2 for leading (LO) and next-to-leading order (NLO) sets as a function of A. From [9].

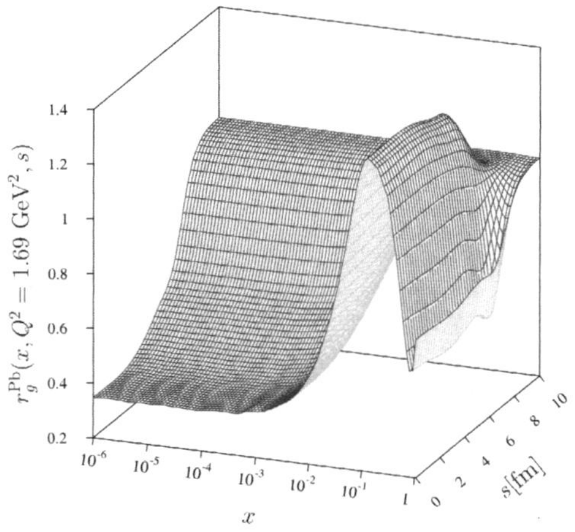

Figure 2. Spatially dependent NLO nuclear modification for gluons in the Pb nucleus from EPS09s as a function of x and the transverse distance $s = |\mathbf{s}|$ at the initial scale Q_0^2. From [9].

3. Results

3.1. Centrality classes

We define the centrality classes in terms of impact parameter intervals which are calculated using the optical Glauber model. For the p+Pb collisions at the LHC with $\sqrt{s_{NN}} = 5.0\,\mathrm{TeV}$ we use $\sigma_{\mathrm{inel}}^{NN} = 70\,\mathrm{mb}$ [12] and, assuming a point-like proton ($T_{\mathrm{pPb}} = T_{\mathrm{Pb}}$), we obtain the centrality classes listed in table 1. For p+Pb collisions with $\sqrt{s_{NN}} = 8.8\,\mathrm{TeV}$ we have used $\sigma_{\mathrm{inel}}^{NN} = 76\,\mathrm{mb}$.

Table 1. The impact parameter intervals and average number of binary collisions for p+Pb collisions with two collision energies for four different centrality classes.

	$\sqrt{s_{NN}} = 5.0\,\mathrm{TeV}$			$\sqrt{s_{NN}} = 8.8\,\mathrm{TeV}$		
	b_1 [fm]	b_2 [fm]	$\langle N_{\mathrm{bin}}\rangle$	b_1 [fm]	b_2 [fm]	$\langle N_{\mathrm{bin}}\rangle$
$0 - 20\,\%$	0.0	3.471	14.24	0.0	3.494	15.44
$20 - 40\,\%$	3.471	4.908	11.41	3.494	4.941	12.31
$40 - 60\,\%$	4.908	6.012	7.663	4.941	6.052	8.165
$60 - 80\,\%$	6.012	6.986	3.680	6.052	7.029	3.810

3.2. Centrality dependent nuclear modification factor

Incorporating the spatially dependent nPDFs into the factorization formula (1) and multiplying this with the nuclear overlap function gives the yield of hard partons k corresponding to a specific impact parameter $\mathbf{b}$ in a collision of nuclei A and B as follows:

$$\begin{aligned} \mathrm{d}N^{AB\to k+X}(\mathbf{b}) = \int \mathrm{d}^2\mathbf{s}\, T_A(\mathbf{s_1})\, T_B(\mathbf{s_2}) \sum_{i,j,X'} r_i^A(x_1, Q^2, \mathbf{s_1}) f_i^N(x_1, Q^2) \\ \otimes\, r_j^B(x_2, Q^2, \mathbf{s_2}) f_j^N(x_2, Q^2) \otimes \mathrm{d}\hat{\sigma}^{ij\to k+X'}, \end{aligned} \tag{7}$$

where $\mathbf{s_1} = \mathbf{s} + \mathbf{b}/2$ and $\mathbf{s_2} = \mathbf{s} - \mathbf{b}/2$. The nuclear effects can be studied by comparing the cross section in nuclear collisions to the same observable in p+p collisions. For this, we define the centrality dependent nuclear modification factor as

$$R^k_{AB}(b_1, b_2) \equiv \frac{\left\langle \frac{\mathrm{d}^2 N^k_{AB}}{\mathrm{d}p_T \mathrm{d}y} \right\rangle_{b_1,b_2}}{\frac{\langle N_{\mathrm{bin}} \rangle_{b_1,b_2}}{\sigma^{NN}_{\mathrm{inel}}} \frac{\mathrm{d}^2\sigma^k_{\mathrm{pp}}}{\mathrm{d}p_T \mathrm{d}y}} = \frac{\int_{b_1}^{b_2} \mathrm{d}^2\mathbf{b} \frac{\mathrm{d}^2 N^k_{AB}(\mathbf{b})}{\mathrm{d}p_T \mathrm{d}y}}{\int_{b_1}^{b_2} \mathrm{d}^2\mathbf{b}\, T_{AB}(\mathbf{b}) \frac{\mathrm{d}^2\sigma^k_{\mathrm{pp}}}{\mathrm{d}p_T \mathrm{d}y}}. \tag{8}$$

Here, N_{bin} is the number of binary collisions and the average is taken over the desired impact parameter interval. Working out the averages yields the expression on the right-hand side. The minimum bias R_{AB} is restored by integrating over the whole impact parameter space.

3.3. Inclusive neutral pion production

The yield of single inclusive neutral pions can be calculated by convoluting the hard parton spectra with the non-perturbative parton-to-hadron fragmentation functions (FFs) $D_{\pi^0/k}(z, Q^2_F)$

$$\mathrm{d}N^{\pi^0}_{AB}(\mathbf{b}) = \sum_k \mathrm{d}N^k_{AB}(\mathbf{b}) \otimes D_{\pi^0/k}(z, Q^2_F), \tag{9}$$

where the k runs over parton flavors and z describes the momentum fraction carried away by the pion. As the FFs are smooth distributions in z, one cannot probe any specific value of x_2 even with fixed p_T and y, but the cross sections get contribution from a wide range of x_2. This is illustrated in Figure 3 showing the normalized inclusive pion production cross section at $p_T = 5\,\mathrm{GeV/c}$ for three rapidities $y = 0, 2, 4$ as a function of x_2. As expected, towards the forward rapidities the kinematic limit is pushed to smaller values of x_2, but the contribution from $x > 10^{-3}$ proves always significant. This is more clearly visible in Figure 4 which illustrates how the cross section builds up when integrating over x_2.

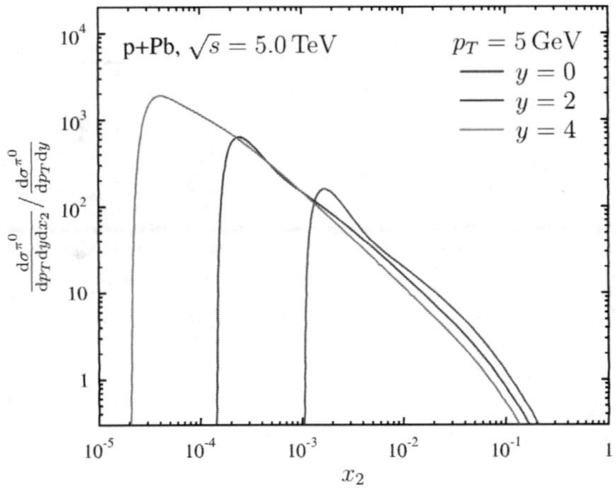

Figure 3. The normalized cross section of π^0 production in p+Pb collisions with $\sqrt{s_{NN}} = 5.0\,\mathrm{TeV}$ for $p_T = 5\,\mathrm{GeV/c}$ and three rapidities, $y = 0$ (green), $y = 2$ (blue), and $y = 4$ (red) as a function of x_2.

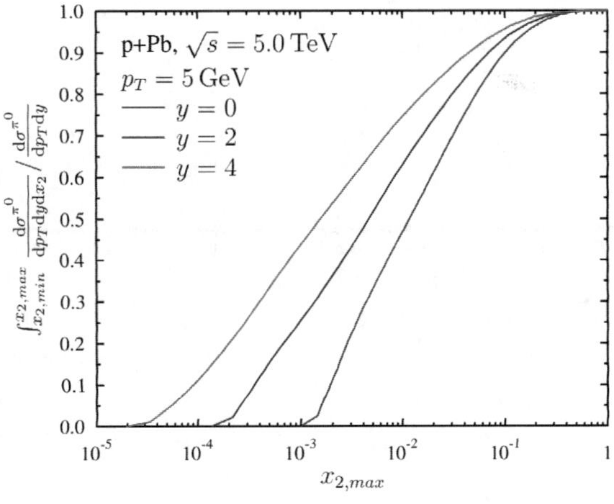

Figure 4. The cumulative cross section of π^0 production as a function of x_2 in p+Pb collisions with $\sqrt{s_{NN}} = 5.0\,\mathrm{TeV}$ for $p_T = 5\,\mathrm{GeV/c}$ and with three rapidities, $y = 0$ (green), $y = 2$ (blue), and $y = 4$ (red).

In Figure 5, we plot our results for the inclusive pion nuclear modification factors in p+Pb collisions with $\sqrt{s_{NN}} = 5.0\,\mathrm{TeV}$ at midrapidity ($y = 0$) for four different centrality classes using the EPS09s nuclear PDFs with different FFs: KKP [13], AKK [14], and fDSS [15]. The calculations are done with `INCNLO` [16, 17, 18, 19], a public NLO code which we have modified to include also the latest FFs and to improve the convergence of the integrals at large y and low p_T with high $\sqrt{s_{NN}}$. All the relevant scales have been set equal to the pion p_T and the uncertainty bands were calculated using the EPS09s error sets. First of all, we notice that even though there are considerable differences in the employed FFs (see e.g. [20]), these differences tend to cancel in $R_{\rm pPb}$. With $p_T < 10\,{\rm GeV/c}$ we see some suppression as the cross sections are sensitive mostly to the small-x region of the nucleus which corresponds to the shadowing in the nPDFs. At larger p_T values, a small enhancement can be observed which follows from the antishadowing in EPS09s. The centrality dependence turns out as expected: in central collisions the nuclear effects are stronger than in peripheral ones. However, as the nuclear modifications are only moderate in the kinematic region considered here, also the centrality dependence is rather mild making it a challenging observable to measure.

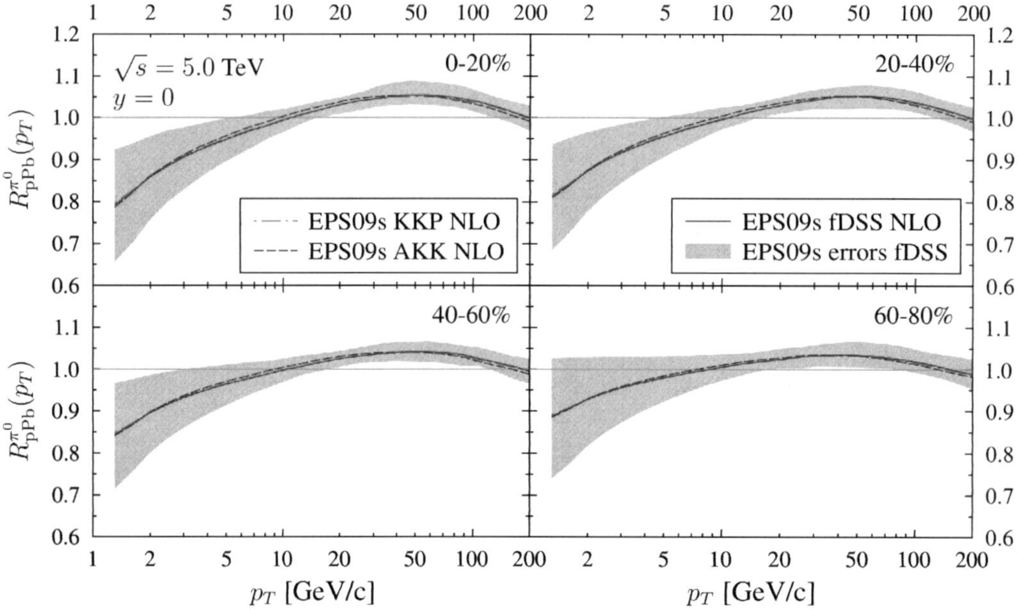

Figure 5. The nuclear modification factor for π^0 production in p+Pb collisions at $\sqrt{s_{NN}} = 5.0\,{\rm TeV}$ and $y = 0$ in different centrality classes (different panels) with EPS09s nPDFs and KKP (red dot-dashed), AKK (blue dashed) and fDSS (green solid) FFs. The uncertainty bands are derived from the EPS09s error sets. Figure is taken from [9].

In order to get further constraints for the nPDF centrality dependence one can consider observables in which the nuclear modifications due to the nPDFs are larger than in the mid-rapidity hadron production. One option is to consider hadron production in forward rapidities as there, according to Figure 3, the cross sections are more sensitive to small x_2 values where the gluon shadowing is stronger. Therefore, Figure 6 presents similar nuclear modification factors as in Figure 5, but now at $y = 4$ (using the fDSS FFs only). The minimum bias result is plotted for comparison. At forward rapidity, the shadowing in the nPDFs indeed generates suppression up to higher values of p_T than at mid-rapidity and the suppression is somewhat stronger. This gives rise to also a slightly more pronounced centrality dependence.

Journal of Physics: Conference Series **589** (2015) 012010 doi:10.1088/1742-6596/589/1/012010

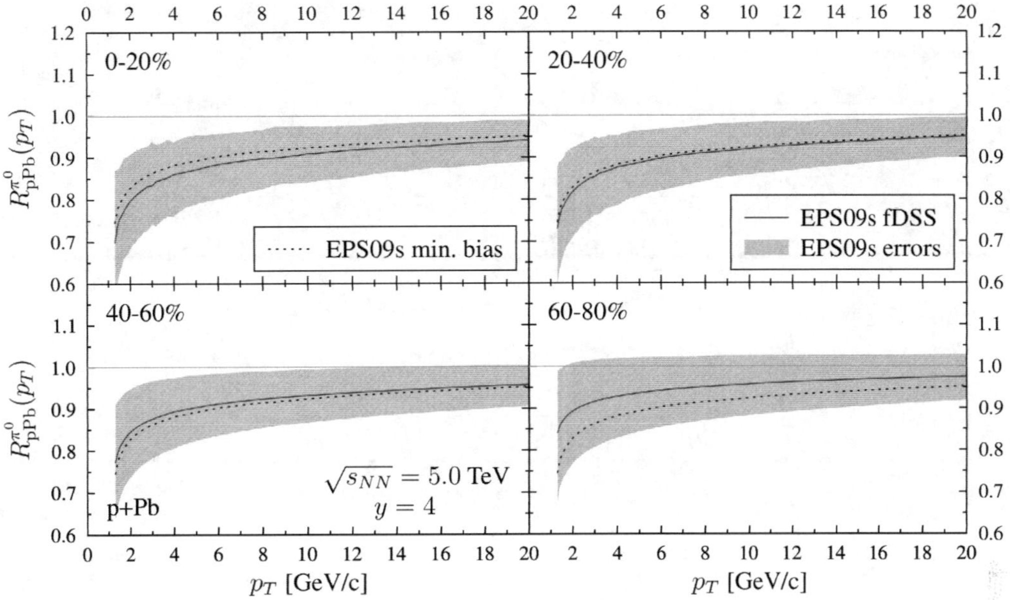

Figure 6. The nuclear modification factor for π^0 production in p+Pb collisions at $\sqrt{s_{NN}} = 5.0\,\text{TeV}$ and $y = 4$ for different centrality classes (different panels) with EPS09s nPDFs and fDSS FFs. The minimum bias result (dotted line) is included in each panel and the uncertainty band is derived from the EPS09s error sets.

3.4. Inclusive prompt photon production

A better way to get sensitivity to small x would be to consider final state particles that are not produced via pure fragmentation process where the convolution with the FFs largely smears the probed partonic x distributions. Low-p_T photons produced directly in the hard scattering of partons would be an ideal candidate to this end. Although it is not possible — experimentally or strictly speaking even in the NLO calculation — to separate these photons from those produced through similar fragmentation mechanism as pions, these two components together, which we refer to as prompt photons, still permit a more direct access to gluon PDFs than the inclusive hadrons. The prompt photon production in minimum bias p+Pb collisions was studied earlier with a similar framework in Ref. [21] and our centrality dependent results at mid-rapidity were published in [22].

To see whether an increased sensitivity to the impact parameter dependence of the nPDFs would be obtainable, the nuclear modification factors for prompt photon production in p+Pb collisions with $\sqrt{s_{NN}} = 5.0\,\text{TeV}$ at $y = 0$ in different centrality classes together with the minimum bias result are plotted in Figure 7. For the calculation of the fragmentation component we have used the BFG II parton-to-photon FFs [23]. When comparing photons in nuclear collisions with the proton-proton collisions one should recall also the presence of an isospin effect which follows from different charge densities between the proton and neutron. This tends to yield some suppression especially at the region which is more sensitive to valence quark distributions even without any nuclear modifications in PDFs. However, from Figure 7 we see that in the kinematic region considered here the isospin effect is negligible. When comparing the result with the corresponding π^0 calculation (Figure 5) we notice that the $R_{\rm pPb}$ at $p_T < 20\,\text{GeV/c}$ is a bit more suppressed for prompt photons, but the difference is rather small. Thus, the centrality dependence turns out to be only slightly larger than that for π^0s at mid-rapidity.

To increase the small-x_2 contribution even further, we consider the prompt photon production

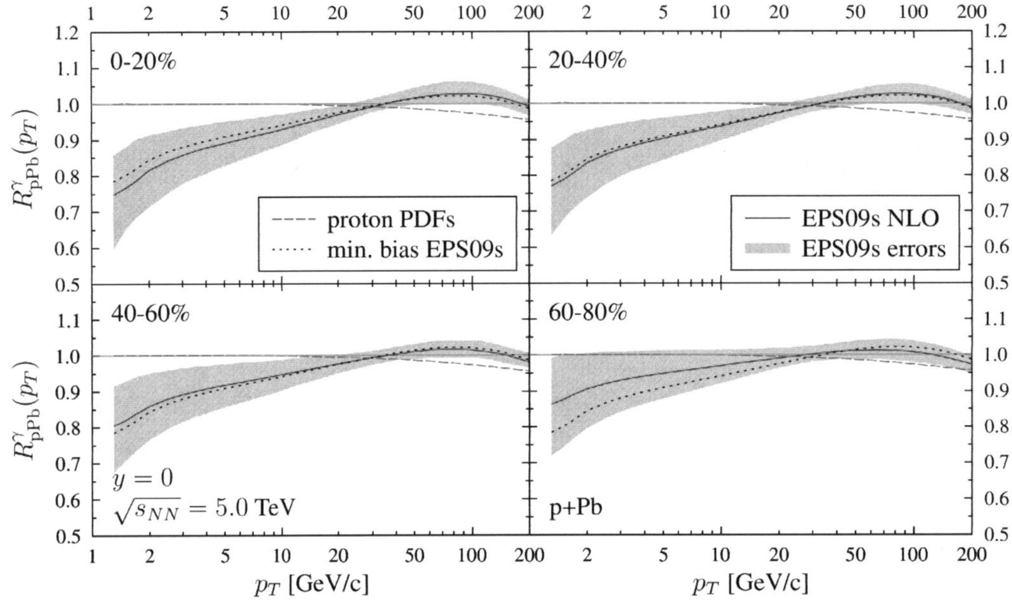

Figure 7. The nuclear modification factor of inclusive prompt photon production in p+Pb collisions at $\sqrt{s_{NN}} = 5.0$ TeV and $y = 0$ for different centrality classes (different panels) with EPS09s nPDFs. The minimum bias result is included to each panel with dotted line, the red dashed line quantifies the isospin effect and the blue uncertainty band is derived from EPS09s error sets. From [22].

also at forwards rapidities. In Figure 8 we show the R^{γ}_{pPb} for p+Pb collisions with $\sqrt{s_{NN}} = 8.8$ TeV at $y = 4.5$ for four centrality classes and for minimum bias collisions. Even in this extreme kinematic region the maximal suppression due to the nPDFs is about 20 percent, except for the very small p_T values. This follows mostly from the rapid DGLAP evolution of the NLO gluon PDFs at small x, and also from the presence of the fragmentation component in prompt photon production which generates contribution also from larger x_2 values.

4. Summary

Using the recently published spatially dependent nuclear PDF set EPS09s, we have calculated nuclear modification factors for p+Pb collisions at the LHC for inclusive pion and prompt photon production in different centrality classes at mid- and forward rapidities in the NLO pQCD framework. The calculations predict larger nuclear effects in central collisions compared to peripheral collisions. The centrality dependence is strongest at small p_T and at forward rapidities, but still the effects are of the order 10 % at most. We notice also that even at the forward rapidities the convolution against the FFs makes these observables less sensitive to small x than could be naively expected. This could be improved by imposing an isolation cut for the prompt photons which would suppress the fragmentation component, or/and by triggering on a heavy quarks as suggested in [24].

Acknowledgments

This work was supported by the Academy of Finland, project 133005. I. H. is supported by Magnus Ehrnrooth foundation and the PANU graduate school.

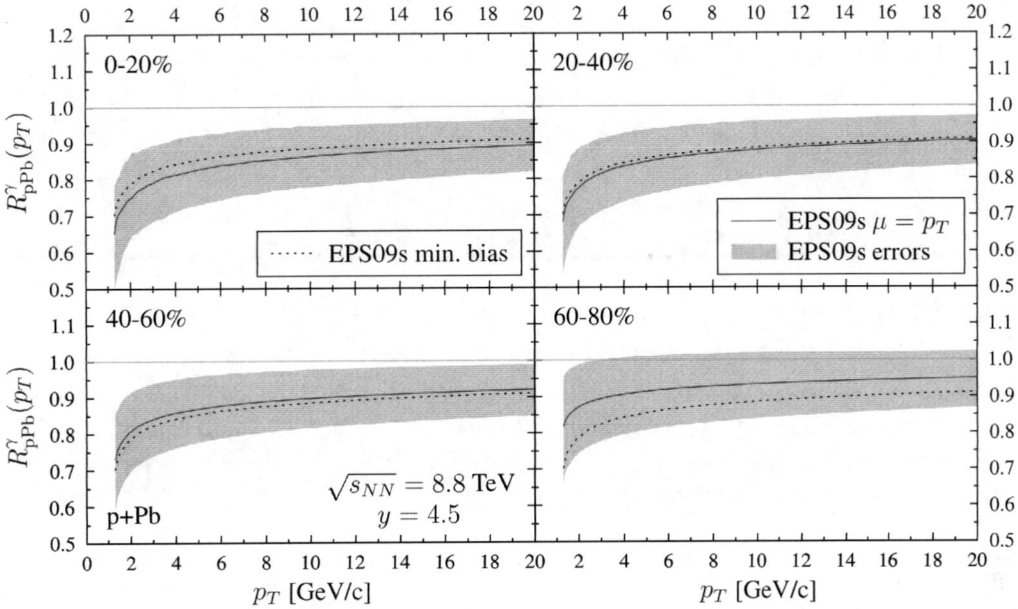

Figure 8. The nuclear modification factor of inclusive prompt photon production in p+Pb collisions at $\sqrt{s_{NN}} = 8.8\,\text{TeV}$ and $y = 4.5$ for different centrality classes (different panels) with EPS09s nPDFs. The minimum bias result is included to each panel with dotted line and the blue uncertainty band is derived from EPS09s error sets.

References

[1] Collins J C, Soper D E and Sterman G F 1988 *Adv.Ser.Direct.High Energy Phys.* **5** 1–91 (*Preprint* hep-ph/0409313)
[2] Brock R *et al.* (CTEQ Collaboration) 1995 *Rev.Mod.Phys.* **67** 157–248
[3] Eskola K J, Paukkunen H and Salgado C A 2009 *JHEP* **04** 065 (*Preprint* 0902.4154)
[4] de Florian D, Sassot R, Zurita P and Stratmann M 2012 *Phys.Rev.* **D85** 074028 (*Preprint* 1112.6324)
[5] Hirai M, Kumano S and Nagai T H 2007 *Phys.Rev.* **C76** 065207 (*Preprint* 0709.3038)
[6] Kovarik K, Jezo T, Kusina A, Olness F, Schienbein I *et al.* 2013 (*Preprint* 1307.3454)
[7] Eskola K J 2013 *Nucl. Phys.* **A 910-911** 163 (*Preprint* 1209.1546)
[8] Paukkunen H 2014 (*Preprint* 1401.2345)
[9] Helenius I, Eskola K J, Honkanen H and Salgado C A 2012 *JHEP* **1207** 073 (*Preprint* 1205.5359)
[10] Albacete J, Armesto N, Baier R, Barnafoldi G, Barrette J *et al.* 2013 *Int.J.Mod.Phys.* **E22** 1330007 (*Preprint* 1301.3395)
[11] Eskola K J, Kolhinen V J and Salgado C A 1999 *Eur.Phys.J.* **C9** 61–68 (*Preprint* hep-ph/9807297)
[12] Antchev G, Aspell P, Atanassov I, Avati V, Baechler J *et al.* 2011 *Europhys.Lett.* **96** 21002 (*Preprint* 1110.1395)
[13] Kniehl B A, Kramer G and Potter B 2000 *Nucl.Phys.* **B582** 514–536 (*Preprint* hep-ph/0010289)
[14] Albino S, Kniehl B A and Kramer G 2008 *Nucl.Phys.* **B803** 42–104 (*Preprint* 0803.2768)
[15] de Florian D, Sassot R and Stratmann M 2007 *Phys.Rev.* **D75** 114010 (*Preprint* hep-ph/0703242)
[16] Aurenche P, Baier R, Fontannaz M and Schiff D 1988 *Nucl.Phys.* **B297** 661
[17] Aversa F, Chiappetta P, Greco M and Guillet J 1989 *Nucl.Phys.* **B327** 105
[18] Aurenche P, Fontannaz M, Guillet J, Kniehl B A and Werlen M 2000 *Eur.Phys.J.* **C13** 347–355 (*Preprint* hep-ph/9910252)
[19] http://lapth.in2p3.fr/PHOX_FAMILY/readme_inc.html
[20] d'Enterria D, Eskola K J, Helenius I and Paukkunen H 2013 (*Preprint* 1311.1415)
[21] Arleo F, Eskola K J, Paukkunen H and Salgado C A 2011 *JHEP* **1104** 055 (*Preprint* 1103.1471)
[22] Helenius I, Eskola K J and Paukkunen H 2013 *JHEP* **1305** 030 (*Preprint* 1302.5580)
[23] Bourhis L, Fontannaz M and Guillet J 1998 *Eur.Phys.J.* **C2** 529–537 (*Preprint* hep-ph/9704447)
[24] Stavreva T, Schienbein I, Arleo F, Kovarik K, Olness F *et al.* 2011 *JHEP* **1101** 152 (*Preprint* 1012.1178)

Journal of Physics: Conference Series **589** (2015) 012011 doi:10.1088/1742-6596/589/1/012011

Corrections and Uncertainties in Energy Loss Model Predictions

W. A. Horowitz

Department of Physics, University of Cape Town, Private Bag X3, Rondebosch 7701, South Africa

E-mail: wa.horowitz@uct.ac.za

Abstract. Energy loss models based on leading order strong-coupling derivations oversuppress high-p_T particles compared to RHIC and LHC data whereas energy loss models based on leading order weak-coupling derivations robustly describe qualitatively known RHIC and LHC results. Currently there are large theoretical uncertainties in both calculations and next-to-leading order corrections will be necessary for 1) a definitive falsification of the strong-coupling calculations and 2) a quantitative comparison of pQCD-based calculations to data.

1. Introduction

The great difficulty currently facing heavy ion physics is the apparent contradiction between the interpretation of low-p_T and high-p_T observables. A consensus has formed in which the distribution of low momentum particles is the result of rapid thermalization followed by nearly ideal hydrodynamic evolution. The best explanation of the early onset of thermalization [1] and nearly ideal fluid flow [2] is the existence of a strongly-coupled fluid best described by the methods of the AdS/CFT correspondence. On the other hand, naive application of the AdS/CFT correspondence to high-p_T probes yield results in contradiction with data. At the same time leading order pQCD predictions predicated on a weakly-coupled plasma weakly coupled to a high momentum probe [3] systematically describe the high-p_T data within a factor of 2 [4]; higher order correction seem likely to lead to an even better description of data [5]. Leading order [6] and sophisticated next-to-leading order [7] calculations based on the same weak-coupling perturbative picture of the plasma, though, yield a thermalization time and a viscosity to entropy ratio an order of magnitude larger than suggested by data. A hybrid strong-weak approach might reconcile these two pictures.

2. Energy Loss in AdS/CFT

2.1. Heavy Quarks

The now well-known leading order analytic energy loss formula for heavy quarks strongly-coupled to a strongly-coupled $\mathcal{N} = 4$ SYM plasma is

$$\frac{dp}{dt} = -\mu\, p, \text{ where } \mu = \frac{\pi\sqrt{\lambda}T^2}{2M_q}. \qquad (1)$$

$\lambda = g^2 N_c$ is the 't Hooft coupling for the theory, T is the temperature of the plasma, and M_q is the mass of the heavy quark [8, 9]. The form of this energy loss is very different from that

Journal of Physics: Conference Series **589** (2015) 012011 doi:10.1088/1742-6596/589/1/012011

found assuming a probe weakly coupled to a weakly-coupled plasma: incoherent Bethe-Heitler bremsstrahlung energy loss [10] goes as

$$\left.\frac{dp}{dt}\right|_{BH} \sim -\frac{T^3}{M_q^2}\,p, \tag{2}$$

but the radiative energy loss in the deep-LPM regime is

$$\left.\frac{dp}{dt}\right|_{LPM} \sim -L\,T^3 \log(p/M_q), \tag{3}$$

where L is the length of the medium through which the heavy quark has passed [11].

Now $\mathcal{N} = 4$ SYM in the $N_c \to \infty$ and λ large and fixed limit is not QCD. However one hopes that the results from AdS/CFT can provide some useful insight into QCD processes. One of the complications of the dissimilarity of the two theories is that there is not a unique, reasonable mapping of the parameters in QCD to those in AdS/CFT. Using a set of these reasonable mappings, one finds a good description of the suppression of heavy quark decay fragments seen by RHIC experiments; see Fig. 1 (a). (A more detailed description of the model and parameters used to compute the figure can be found in [12].)

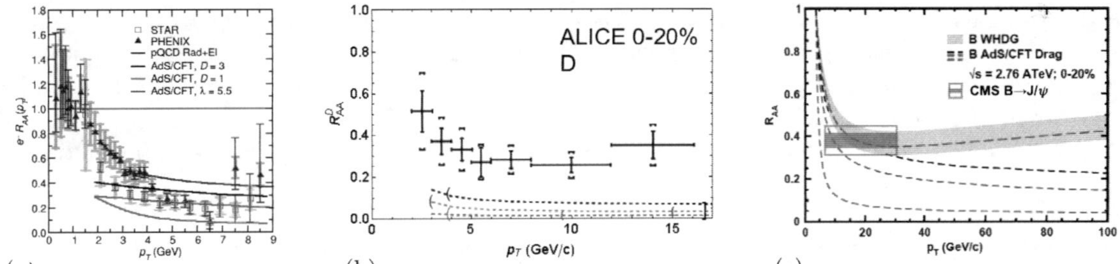

(a) (b) (c)

Figure 1: (a) AdS/CFT (and pQCD) predictions [12] for non-photonic electron decay products of heavy c and b quarks at RHIC [13, 14] and (b) D meson and (c) B meson suppression predictions from AdS/CFT [15] at LHC [16, 17]

Using this exact same set of mappings one may make predictions for D and B meson suppression at LHC, shown in Fig. 1 (b) and (c). As seen in (c), given the current uncertainties in the theoretical predictions and experimental measurements, the B meson predictions are consistent with data. The D mesons, however, as shown in (b), are falsifiably oversuppressed compared to ALICE data.

Previous calculations [12] included estimates of a "speed limit" for the applicability of the heavy quark drag calculations. Several independent lines of reasoning [18, 19] imply that the formalism does not apply to heavy quarks propagating faster than

$$\gamma \lesssim \gamma_{critical} = \left(1 + \frac{2M_q}{\sqrt{\lambda}T}\right)^2 \sim \frac{4M_q^2}{\lambda T^2}. \tag{4}$$

This speed limit is parametrically smaller than an estimate for the momenta at which the fluctuations in momentum loss become important [18],

$$\gamma_{fluc} \sim \frac{M_q^2}{4T^2}; \tag{5}$$

however, it turns out that numerically this latter speed limit is reached first. Efforts are underway to quantify the importance of momentum fluctuations in the suppression of high momentum heavy quarks.

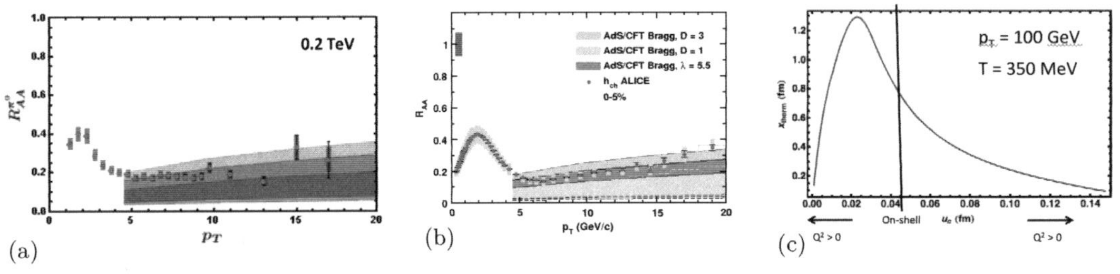

Figure 2: Predictions from a simple Bragg peak model for AdS/CFT light flavor energy loss [22] compared (a) to PHENIX data at RHIC [23] and (b) ALICE data at LHC [24]. (c) The stopping distance in AdS/CFT for light flavors depends strongly on the initial conditions.

2.2. Light Flavors

A critical test of any energy loss formalism is a simultaneous description of both heavy and light flavor suppression. The original calculations of light flavor energy loss require difficult numerics as the endpoints of the string are allowed to dynamically fall in the 5th dimension [20]. More recent work using an alternative setup yields a simple analytic solution [21]. However, it is not yet entirely clear what is the most appropriate setup in the AdS space to model light flavor energy loss. In the following, we will attempt to infer the physical consequences of the original light flavor energy loss calculation.

One of the first observations of the original setup was a generic Bragg peak in the energy loss such that the maximum stopping distance scaled as

$$\Delta x_{max} \sim \left(\frac{E}{\sqrt{\lambda}T}\right)^{1/3}\frac{1}{T}. \tag{6}$$

One can create a very naive energy loss model based on this maximum stopping distance [22]: assume that any light flavor created with $L < x_{max}$ gets out of the plasma unaltered while flavors created with $L > x_{max}$ are completely absorbed. There are large uncertainties in this model; in addition to the usual unknown mapping from QCD to AdS/CFT, one also does not know which single value of T to plug into Eq. (6). A maximal uncertainty band can be created by taking two extreme values for T: 1) the temperature at the point of creation at the moment of thermalization or 2) the transition temperature between the deconfined and confined phases of QCD matter. The predictions resulting from these two extremes are shown as a band in Fig. 2 (a) and (b); the data at RHIC and LHC fall within the very large theoretical uncertainties.

Since the most naive calculation is not obviously falsified by the data, it is worth pursuing a more precise theoretical model. Preliminary investigations [25] show that results depend very sensitively on the "jet" prescription chosen, to the extent that one can even make the Bragg peak in the energy loss appear and disappear. Additional, large sensitivity comes from the precise initial string profile propagated from early times in the collision; see Fig. 2 (c). It turns out that very little of the $2 \times \infty$ dimensions of the space of initial conditions has been explored. We will return to these issues later in this proceedings.

3. pQCD Energy Loss

One may also choose to determine the consequences of alternative picture of a weakly-coupled plasma weakly coupled to a high momentum probe. Jacksonian intuition suggests that at the GeV scale radiative energy loss dominates over collisional. However detailed calculations [26] show that the elastic energy loss is of the same order of magnitude as inelastic at the energy

Journal of Physics: Conference Series **589** (2015) 012011 doi:10.1088/1742-6596/589/1/012011

regimes applicable at RHIC and LHC; see Fig. 3. The PHENIX collaboration performed [27] a

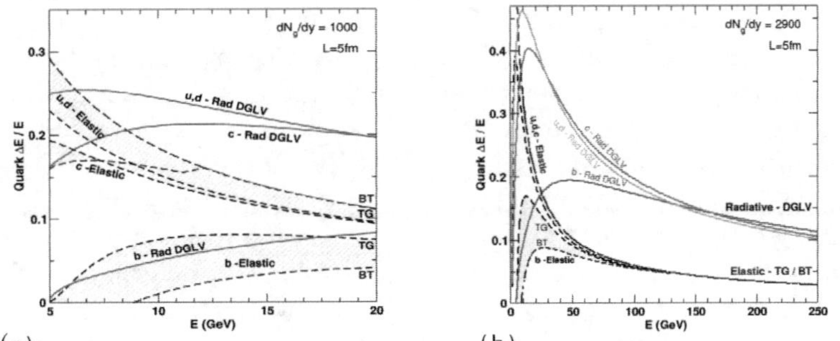

(a) (b)

Figure 3: Comparison of magnitude of radiative and collisional energy loss in pQCD for (**??**) RHIC and (b) LHC [26]. Due to the LPM effect, elastic energy loss is the same order of magnitude as inelastic for all particle energies.

rigorous statistical analysis of the WHDG energy loss model [3] that incorporates both radiative and collisional energy loss in a reasonable geometric background that extracted the one free parameter in the calculation: the proportionality constant between the participant density and the number density of the color deconfined medium produced in heavy ion collisions. The value found by PHENIX corresponds to a central gluon rapidity density of $dN_g/dy = 1400^{+200}_{-375}$. Keeping the proportionality constant fixed, varying the medium density at different centralities and center of mass energies only by the measured multiplicities, the model robustly describes qualitatively a wealth of high-p_T observables; see Fig. 4, in which the theoretical uncertainty band is due only to the 1-σ range of values from the PHENIX proportionality constant extraction.

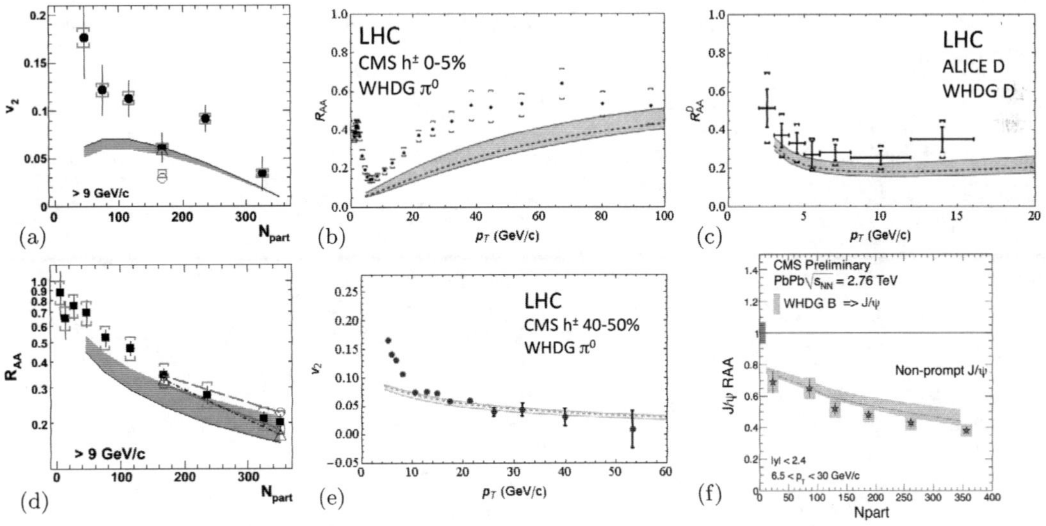

(a) (b) (c) (d) (e) (f)

Figure 4: Constrained zero parameter WHDG predictions compared to data for (a) $v_2^{\pi^0}(N_{part})$ at RHIC [23, 3], (b) 0-5% centrality $R_{AA}(p_T)$ for light flavors at LHC [28, 29], (c) $R_{AA}^{D}(p_T)$ at 0-20% centrality at LHC [16, 22], (d) $R_{AA}^{\pi^0}(N_{part})$ at RHIC [23, 3], (e) $v_2(p_T)$ at LHC for light flavors at 40-50% centrality [30, 22], and (f) $R_{AA}^{B\to J\psi}(N_{part})$ at LHC [31, 15].

There are a very large number of sources of theoretical uncertainty not shown in the results above. Some of these sources of uncertainty include higher order contributions in: coupling

α_s; collinearity, or k_T/xE, where k_T is the radiated gluon's perpendicular momentum, and xE is the fraction of the leading parton's initial energy carried away by the emitted gluon; softness, x; quark mass to energy of the leading heavy quark, M_q/E; and opacity, the ratio of the mean free path to the pathlength, λ_{mfp}/L. An early attempt [32] to estimate the sensitivity of the calculation to higher order contributions in α_s varied the value of α_s from 0.2 to 0.4 and found a strong dependence of the suppression on the value of α_s chosen, not surprisingly as $dp/dt_{coll} \sim \alpha_s^2$ and $dp/dt_{rad} \sim \alpha_s^3$, although the amount of dependence absorbed by reevaluating the proportionality constant as α_s was varied was never explored. More recent work with a running coupling ansatz found in fact a better agreement with pion suppression as a function of p_T than the fixed coupling calculation [5]. Others showed that [33] pQCD calculations rather significantly violate the assumption of collinearity at RHIC and LHC energies and are very sensitive to the treatment of wide angle radiation. Predictions appear stable, i.e. not sensitive, once the proportionality constant is fixed for a given prescription for the treatment of the wide angle radiation [29]. However the inferred properties of the medium depend on the proportionality constant, which may vary by a factor of 3 due to the uncertainty in the treatment of wide angle radiation [33]; quantitative information regarding the medium therefore requires a detailed understanding of the higher order corrections in collinearity.

4. Discussion and Conclusions

How can we move forward to resolve the seeming contradiction between the weakly- and strongly-coupled pictures? How can we narrow down the theoretical uncertainties due to 1) the jet definition prescription and 2) the initial conditions in AdS space setup? We hope to address both of these issues simultaneously with a hybrid weak-strong energy loss model. Perhaps the early energy loss evolution is dominated by weak-coupling physics but later evolution, as the medium cools, is dominated by strong-coupling physics. We are in the process of creating a model that interfaces these two regimes by matching the finite time energy-momentum tensor of a high momentum colored object created in a heavy ion collision as calculated in pQCD to that calculated in AdS/CFT.

5. Acknowledgments

The author wishes to thank the South African National Research Foundation and SA-CERN for support.

6. References

[1] Chesler P M and Yaffe L G 2011 *Phys.Rev.Lett.* **106** 021601 (*Preprint* 1011.3562)
[2] Kovtun P, Son D and Starinets A 2005 *Phys.Rev.Lett.* **94** 111601 (*Preprint* hep-th/0405231)
[3] Wicks S, Horowitz W, Djordjevic M and Gyulassy M 2007 *Nucl.Phys.* **A784** 426–442 (*Preprint* nucl-th/0512076)
[4] Horowitz W 2013 *Nucl.Phys.A904-905* **2013** 186c–193c (*Preprint* 1210.8330)
[5] Buzzatti A and Gyulassy M 2013 *Nucl.Phys.A904-905* **2013** 779c–782c (*Preprint* 1210.6417)
[6] Danielewicz P and Gyulassy M 1985 *Phys.Rev.* **D31** 53–62
[7] Chen J W, Deng J, Dong H and Wang Q 2013 *Phys.Rev.* **C87** 024910 (*Preprint* 1107.0522)
[8] Gubser S S 2006 *Phys.Rev.* **D74** 126005 (*Preprint* hep-th/0605182)
[9] Herzog C, Karch A, Kovtun P, Kozcaz C and Yaffe L 2006 *JHEP* **0607** 013 (*Preprint* hep-th/0605158)
[10] Bethe H and Heitler W 1934 *Proc.Roy.Soc.Lond.* **A146** 83–112
[11] Djordjevic M and Gyulassy M 2004 *Nucl.Phys.* **A733** 265–298 (*Preprint* nucl-th/0310076)
[12] Horowitz W and Gyulassy M 2008 *Phys.Lett.* **B666** 320–323 (*Preprint* 0706.2336)
[13] Dion A (PHENIX collaboration) 2009 *Nucl.Phys.* **A830** 765C–768C (*Preprint* 0907.4749)
[14] Bielcik J (STAR Collaboration) 2006 *Nucl.Phys.* **A774** 697–700 (*Preprint* nucl-ex/0511005)
[15] Horowitz W 2012 *AIP Conf.Proc.* **1441** 889–891 (*Preprint* 1108.5876)
[16] Abelev B *et al.* (ALICE Collaboration) 2012 *JHEP* **1209** 112 (*Preprint* 1203.2160)
[17] Chatrchyan S *et al.* (CMS Collaboration) 2012 *JHEP* **1205** 063 (*Preprint* 1201.5069)
[18] Gubser S S 2008 *Nucl.Phys.* **B790** 175–199 (*Preprint* hep-th/0612143)

Journal of Physics: Conference Series **589** (2015) 012011 doi:10.1088/1742-6596/589/1/012011

[19] Casalderrey-Solana J and Teaney D 2007 *JHEP* **0704** 039 (*Preprint* hep-th/0701123)
[20] Chesler P M, Jensen K, Karch A and Yaffe L G 2009 *Phys.Rev.* **D79** 125015 (*Preprint* 0810.1985)
[21] Ficnar A and Gubser S S 2014 *Phys.Rev.* **D89** 026002 (*Preprint* 1306.6648)
[22] Horowitz W and Gyulassy M 2011 *J.Phys.* **G38** 124114 (*Preprint* 1107.2136)
[23] Adare A *et al.* (PHENIX Collaboration) 2010 *Phys.Rev.Lett.* **105** 142301 (*Preprint* 1006.3740)
[24] Aamodt K *et al.* (ALICE Collaboration) 2011 *Phys.Lett.* **B696** 30–39 (*Preprint* 1012.1004)
[25] Morad R and Horowitz W A *in preparation*
[26] Horowitz W A 2010 (*Preprint* 1011.4316)
[27] Adare A *et al.* (PHENIX Collaboration) 2008 *Phys.Rev.* **C77** 064907 (*Preprint* 0801.1665)
[28] Chatrchyan S *et al.* (CMS Collaboration) 2012 *Eur.Phys.J.* **C72** 1945 (*Preprint* 1202.2554)
[29] Horowitz W and Gyulassy M 2011 *Nucl.Phys.* **A872** 265–285 (*Preprint* 1104.4958)
[30] Chatrchyan S *et al.* (CMS Collaboration) 2012 *Phys.Rev.Lett.* **109** 022301 (*Preprint* 1204.1850)
[31] Mironov C (CMS Collaboration) 2013 *Nucl.Phys.* **A904-905** 194c–201c
[32] Wicks S 2008
[33] Horowitz W and Cole B 2010 *Phys.Rev.* **C81** 024909 (*Preprint* 0910.1823)

Journal of Physics: Conference Series **589** (2015) 012012 doi:10.1088/1742-6596/589/1/012012

Parton density constraints from massive vector boson production at the LHC

Michael Klasen

Institut für Theoretische Physik, Westfälische Wilhelms-Universität Münster, Wilhelm-Klemm-Straße 9, D-48149 Münster, Germany

E-mail: michael.klasen@uni-muenster.de

Abstract. We demonstrate that not only the production of virtual photons decaying into low-mass lepton pairs, but also the one of weak bosons at large transverse momenta is dominated by quark-gluon scattering. Measurements of these processes at the LHC can therefore provide useful constraints on the parton densities in the proton, in particular the one of the gluon, and their nuclear modifications.

1. Introduction

Parton distribution functions (PDFs) in protons and nuclei are traditionally determined in deep-inelastic electron scattering. While the (valence) quark densities can in this way well be determined, the gluon and sea quark densities enter only at higher order in perturbation theory, so that they are less well determined, in particular at very small and large momentum fractions x. Therefore, constraints from other hard scattering processes like inclusive hadron or jet production must be added in global analyses [1, 2, 3]. In nuclear collisions, the free proton PDFs are modified by the presence of the surrounding nucleons. From small to large x, shadowing, antishadowing and EMC effects, Fermi motion and isospin modifications occur. These effects are currently much less well determined than the PDFs in free protons and are parametrised by various groups [4, 5, 6, 7].

Prompt photon production at sufficiently large transverse momentum p_T is known to be dominated by quark-gluon scattering, the so-called QCD Compton process, rather than quark-antiquark scattering. It can therefore in principle provide additional constraints on the gluon density and its nuclear modifications [8, 9]. However, fragmentation contributions lead to additional uncertainties [10, 11] and must be reduced by applying isolation criteria [12]. These difficulties do not occur with slightly virtual photons decaying into lepton pairs with small invariant mass, which are still dominantly produced in the QCD Compton process and can thus serve as a surrogate for prompt-photon production [13, 14]. As we have recently shown and will describe in this presentation, even Z and W boson production can serve to constrain the gluon PDFs in the proton at large x, if the transverse momentum of the produced vector boson is much larger than its mass [15], while nuclear modification effects can be constrained from low-mass lepton pair production in proton-ion collisions at the LHC [16].

Journal of Physics: Conference Series **589** (2015) 012012 doi:10.1088/1742-6596/589/1/012012

2. Massive vector boson production at the LHC

The production of W and Z bosons at the LHC with a centre-of-mass energy of $\sqrt{s} = 7$ TeV has recently been measured by the ATLAS [17] and CMS [18] collaborations out to p_T values of 600 GeV. The corresponding distribution for Z bosons is shown in Fig. 1 (top). While at $p_T > 65$ GeV our perturbative calculation at next-to-leading order (NLO) is sufficient, this calculation diverges at low p_T due to large logarithms and must be resummed, which we do up to next-to-leading logarithmic accuracy. As one can see, the theoretical predictions then agree very nicely with the data also at small p_T. Similar results have been obtained for the production of W bosons [15].

Intuitively, one might expect massive Z and W bosons to be produced dominantly through the Drell-Yan process, i.e. in quark-antiquark annihilation, where a large fraction of longitudinal momentum is transferred from the proton to the parton. At the LHC, this is only true at small transverse momenta of the vector boson, as can be seen from Fig. 1 (bottom). When $p_T > M_{W,Z}$, the mass of the vector boson becomes less relevant, so that the QCD Compton process eventually takes over as it does for real and virtual photon production.

3. Parton densities in the proton

The values of x probed by the current data sets, as estimated by $x_T = 2p_T/\sqrt{s}$ at central rapidities, extend only to small and intermediate values of 0.01 ... 0.15, where the quark and gluon densities are relatively well constrained.

In Fig. 2 (top) we show that this changes dramatically as higher values of p_T and x are reached. The PDF uncertainty as estimated by the CT10 error sets (yellow band) then easily exceeds the theoretical uncertainty from scale variations at NLO (red lines). In addition, other PDF parametrisations can fall outside the CT10 error band, indicating that the latter has been underestimated due to the theoretical ansatz at the starting scale or due to the selection of data sets entering the global analysis. Reaching higher values of p_T of course requires higher luminosity, which will however soon become available at centre-of-mass energies of up to $\sqrt{s} = 14$ TeV. At this energy, the x-axes of our figures can simply be rescaled by a factor of two.

Alternatively, large values of x can be reached in the forward or backward rapidity regions. As we show in Fig. 2 (bottom), the PDF uncertainties then exceed those of the scale variations already at currently accessible p_T values and can easily reach 100%.

4. Nuclear modifications of parton densities

Nuclear modifications of parton densities can be determined in ratios of cross sections of proton-ion over proton-proton collisions. These offer the advantage that theoretical uncertainties from scale variations and free proton PDFs cancel out to a large extent in the numerator and denominator, as do most experimental systematic uncertainties.

In 2013, the LHC collided protons with lead ions at a centre-of-mass energy of $\sqrt{s} = 5.02$ TeV with beams circulating in both directions. Reference data were also taken in proton-proton collisions at the same energy. In Fig. 3 we show that the cross section ratios for low-mass muon pairs produced in the forward direction and detected, e.g., with the ALICE muon spectrometer, are very sensitive to nuclear modification effects. If the protons circulate towards the muon spectrometer, the PDFs in the lead ion are produced at relatively low values of x, i.e. in the shadowing region (Fig. 3, top). Only at relatively large values of $p_T > 70$ GeV the antishadowing region is reached, and the exact transition point depends strongly on the nuclear

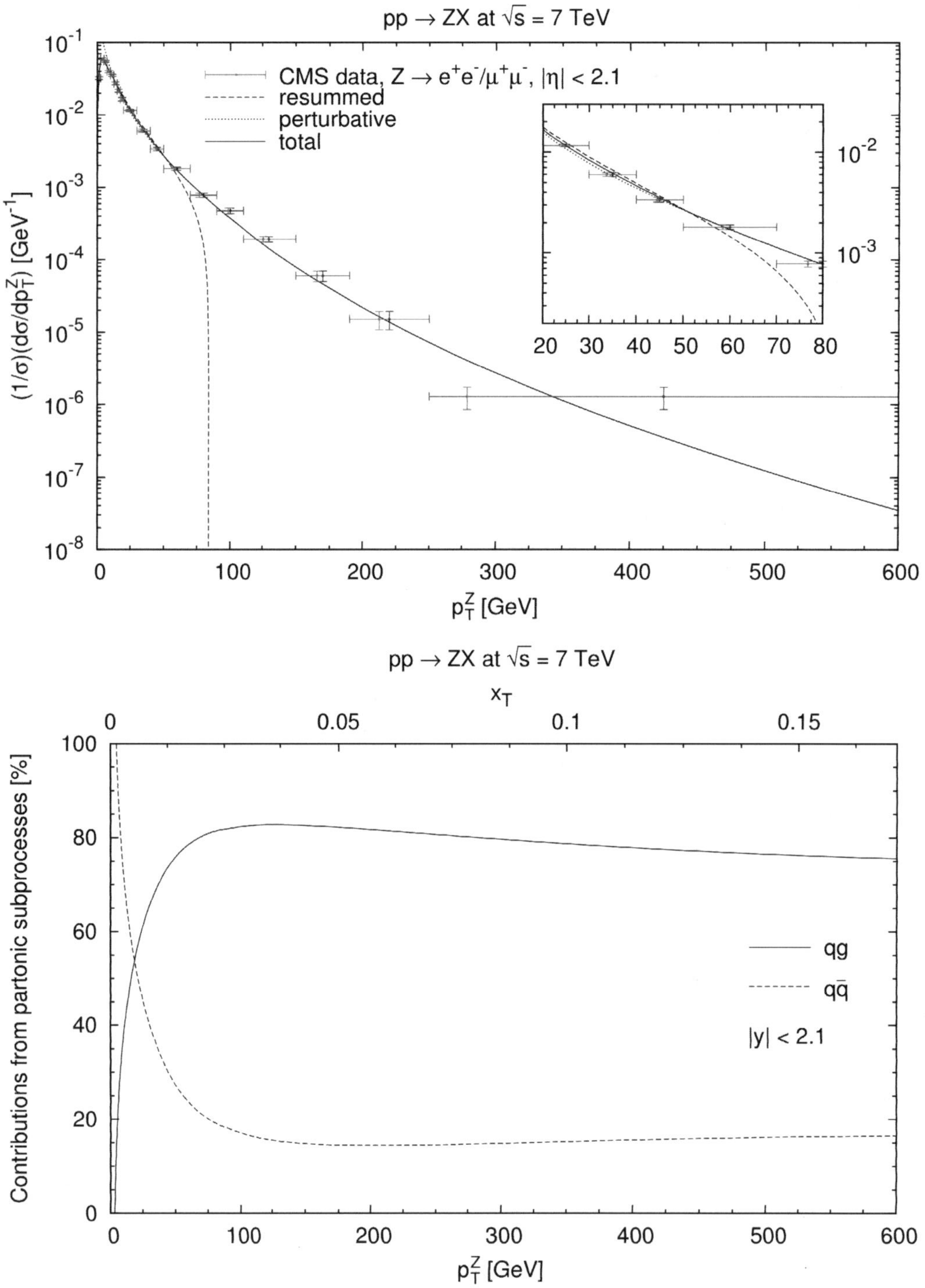

Figure 1. Transverse-momentum distribution of Z bosons produced in pp collisions at the LHC, compared to recent CMS data (top), and its relative contributions from partonic subprocesses (bottom).

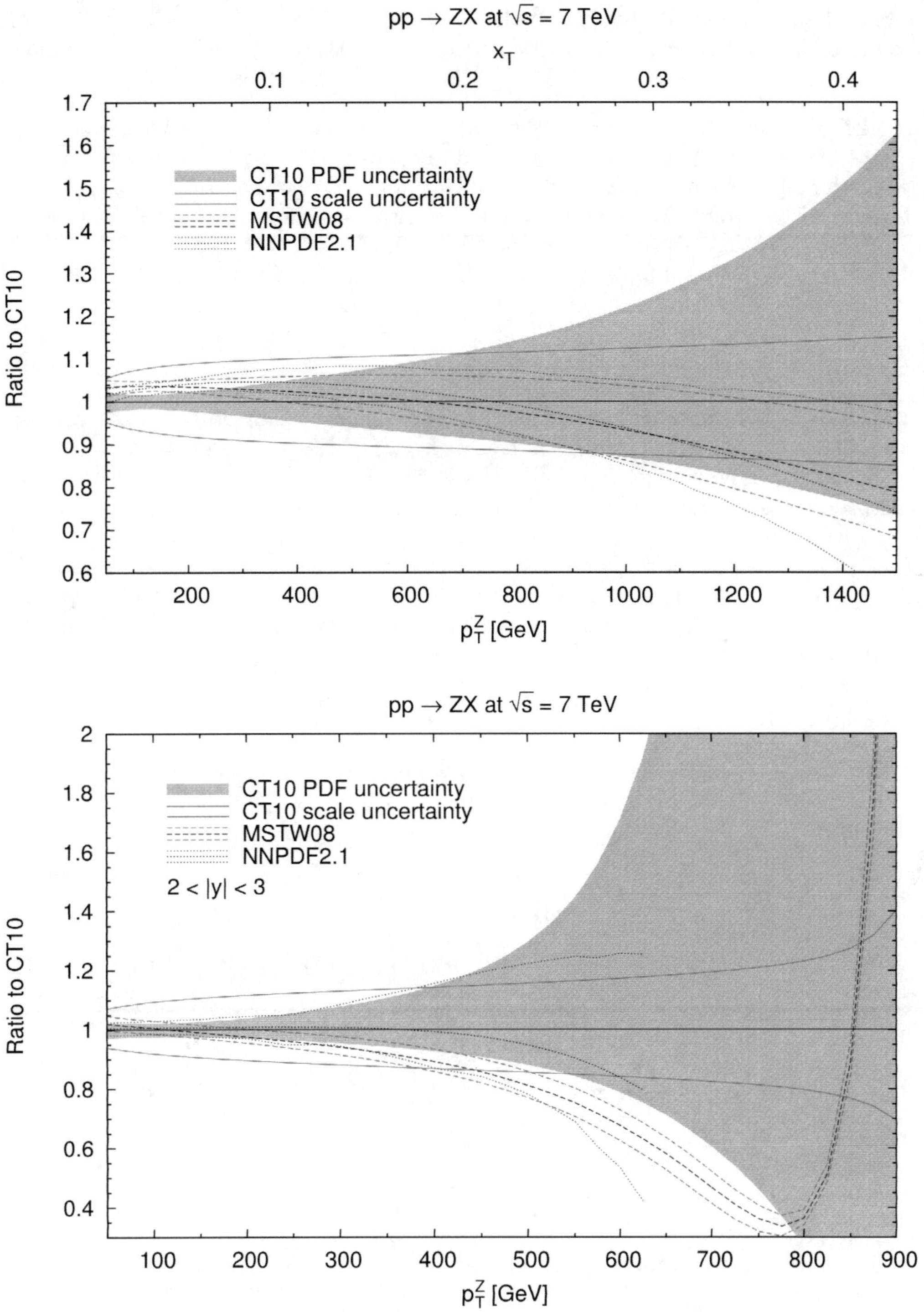

Figure 2. Relative PDF und scale uncertainties of p_T-distributions for Z bosons produced in pp collisions with $\sqrt{s} = 7$ TeV at central (top) and forward (bottom) rapidity.

PDF parametrisation. The uncertainty in the shadowing effect is $\pm 10\%$ at low p_T according to EPS09 and reaches -20% in the nCTEQ prediction, demonstrating again the strong bias of the theoretical ansatz at the starting scale and of the selection of data sets in the global analysis.

When the lead ions circulate towards the muon spectrometer (Fig. 3, bottom), large x values are probed. There, one has to take into account the isospin effect (green curve) and the EMC effect, which for EPS09 is predicted for $p_T > 40$ GeV and $x > 0.4$. In the other parametrisations the antishadowing region extends over almost the entire p_T range shown. Similar studies can be performed with electron-positron pairs in the central detectors of ALICE, ATLAS and CMS. With the latter two general purpose experiments, also much higher values of p_T may eventually be reached.

5. Conclusion

In conclusion, not only low-mass lepton pairs can serve as important probes of the gluon density in the proton, in particular at large x, but also the weak W and Z bosons can play this role despite their mass, provided that their transverse momenta are sufficiently large for the QCD Compton process to become dominant. With low-mass lepton pairs, nuclear modifications of parton densities can in principle be studied in all relevant regimes, provided that sufficient beam time is allocated to proton-ion runs at the LHC to reach luminosities at least in the pb range. Measuring ratios of cross sections then allows for large cancellations of theoretical and experimental uncertainties, leading to considerable improvements of our current knowledge of parton distributions in bound protons.

Acknowledgments

The author thanks the organisers of this conference for the kind invitation to this very stimulating workshop and M. Brandt, C. Klein-Bösing, F. König and J. Wessels for their collaboration.

References

[1] J. Gao, M. Guzzi, J. Huston, H. -L. Lai, Z. Li, P. Nadolsky, J. Pumplin and D. Stump *et al.*, arXiv:1302.6246 [hep-ph].
[2] A. D. Martin, W. J. Stirling, R. S. Thorne and G. Watt, Eur. Phys. J. C **63** (2009) 189 [arXiv:0901.0002 [hep-ph]].
[3] R. D. Ball, V. Bertone, F. Cerutti, L. Del Debbio, S. Forte, A. Guffanti, J. I. Latorre and J. Rojo *et al.*, Nucl. Phys. B **849** (2011) 296 [arXiv:1101.1300 [hep-ph]].
[4] K. J. Eskola, H. Paukkunen and C. A. Salgado, JHEP **0904** (2009) 065 [arXiv:0902.4154 [hep-ph]].
[5] M. Hirai, S. Kumano and T. -H. Nagai, Phys. Rev. C **76** (2007) 065207 [arXiv:0709.3038 [hep-ph]].
[6] D. de Florian, R. Sassot, P. Zurita and M. Stratmann, Phys. Rev. D **85** (2012) 074028 [arXiv:1112.6324 [hep-ph]].
[7] I. Schienbein, J. Y. Yu, K. Kovarik, C. Keppel, J. G. Morfin, F. Olness and J. F. Owens, Phys. Rev. D **80** (2009) 094004 [arXiv:0907.2357 [hep-ph]].
[8] F. Arleo and T. Gousset, Phys. Lett. B **660** (2008) 181 [arXiv:0707.2944 [hep-ph]].
[9] F. Arleo, K. J. Eskola, H. Paukkunen and C. A. Salgado, JHEP **1104** (2011) 055 [arXiv:1103.1471 [hep-ph]].
[10] L. Bourhis, M. Fontannaz and J. P. Guillet, Eur. Phys. J. C **2** (1998) 529 [hep-ph/9704447].
[11] M. Klasen, C. Klein-Bösing, F. König and J. P. Wessels, JHEP **1310** (2013) 119 [arXiv:1307.7034].
[12] S. Catani, M. Fontannaz, J. P. .Guillet and E. Pilon, JHEP **1309** (2013) 007 [arXiv:1306.6498 [hep-ph]].
[13] E. L. Berger, L. E. Gordon and M. Klasen, Phys. Rev. D **58** (1998) 074012 [hep-ph/9803387].
[14] E. L. Berger, L. E. Gordon and M. Klasen, Phys. Rev. D **62** (2000) 014014 [hep-ph/9909446].
[15] M. Klasen and M. Brandt, Phys. Rev. D **88** (2013) 054002 [arXiv:1305.5677 [hep-ph]].
[16] M. Brandt, M. Klasen and F. König, arXiv:1401.6817 [hep-ph].
[17] G. Aad *et al.* [ATLAS Collaboration], Phys. Rev. D **85** (2012) 012005 [arXiv:1108.6308 [hep-ex]].
[18] S. Chatrchyan *et al.* [CMS Collaboration], Phys. Rev. D **85** (2012) 032002 [arXiv:1110.4973 [hep-ex]].

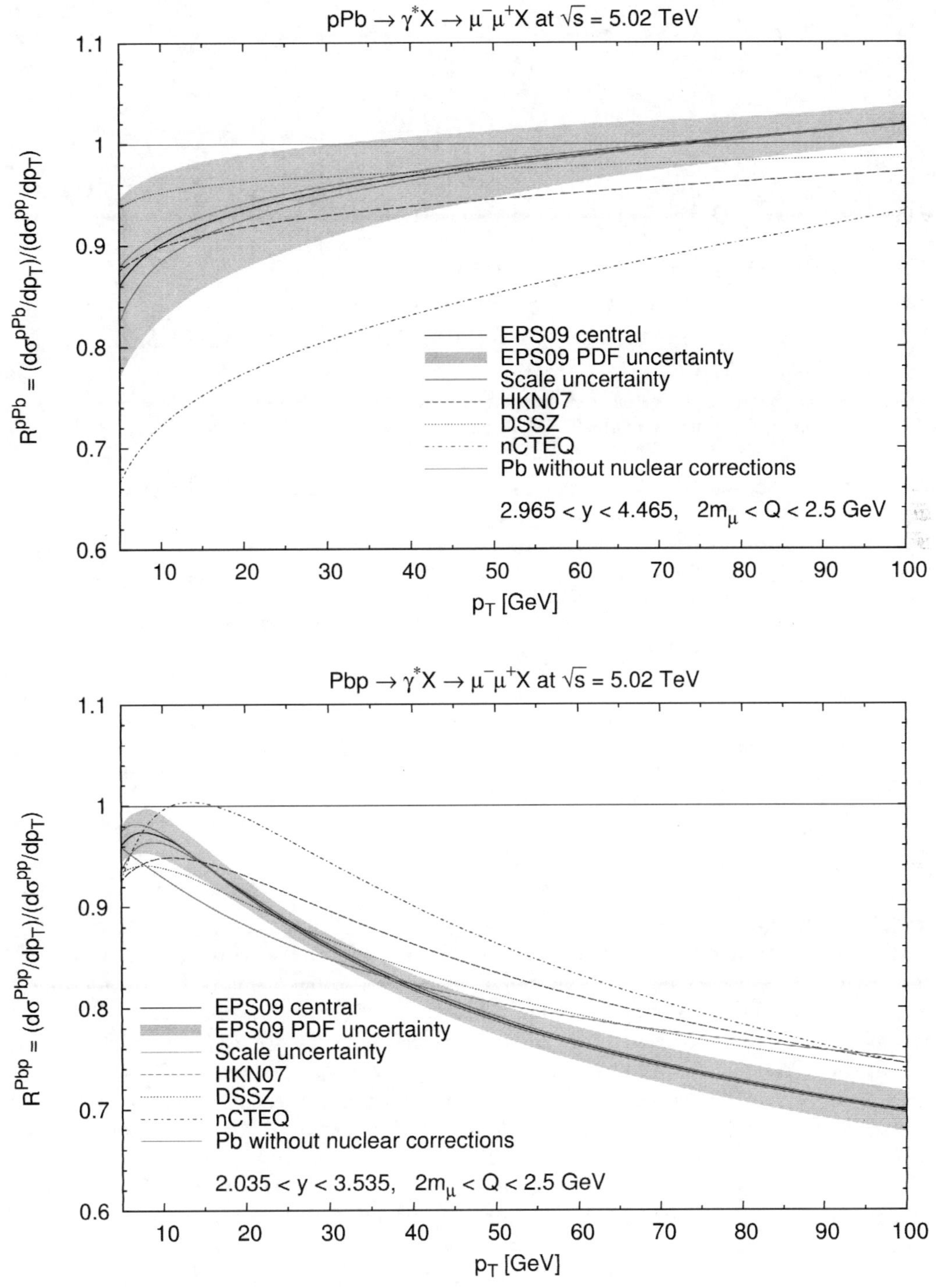

Figure 3. Nuclear modification factors as a function of p_T of low-mass muon pairs produced in proton-lead (top) and lead-proton (bottom) collisions with $\sqrt{s} = 5.02$ TeV at forward rapidity.

Journal of Physics: Conference Series **589** (2015) 012013 doi:10.1088/1742-6596/589/1/012013

High-p_T hadrons in pA and AA collisions: Impact of energy conservation

B.Z. Kopeliovich[1], **J. Nemchik**[2,3], **I.K. Potashnikova**[1] and **Ivàn Schmidt**[1]

[1]Departamento de Física, Universidad Técnica Federico Santa María;
Centro Científico-Tecnológico de Valparaíso, Avda. España 1680, Valparaíso, Chile
[2] Czech Technical University in Prague, FNSPE, Břehová 7, 11519 Prague, Czech Republic
[3] Institute of Experimental Physics SAS, Watsonova 47, 04001 Košice, Slovakia

E-mail: `boris.kopeliovich@usm.cl`

Abstract. The Cronin effect, which is nuclear enhancement of high-p_T hadron production in pA collisions was successfully predicted prior the measurements at RHIC and LHC. The restrictions imposed by energy conservation lead to spectacular effects. Energy deficit becomes an issue for hadron production in pA collisions at large x_L and/or large x_T towards the kinematic bounds $x_{L,T} = 1$. It leads to a suppression, which has been indeed observed for hadrons produced at forward rapidities and large p_T. Intensive energy dissipation via gluon radiation by a highly virtual parton produced with large p_T, makes impossible this process to continue long. Color neutralization and creation of a colorless dipole must occur promptly. When this happens inside a hot medium created in AA collisions, attenuation of dipoles, rather than induced energy loss, becomes a dominant mechanism for suppression of high-p_T hadrons.

1. pA collisions: Cronin effect

High-p_T hadrons can be produced coherently from multiple interactions in nuclei at very high energies (LHC), but not at low energies of fixed target experiments. Correspondingly, the mechanisms for the Cronin enhancement are different. Within the parton model high-p_T hadron production results from hard scattering of the partons, which belong to the colliding hadrons, $1 + 2 \to 3 + 4$. This process looks quite differently in the rest frame of the target, where the produced partons 3 and 4 pre-exist as a high-p_T fluctuation inside the incident hadron. Such a duality in the interpretation of the process is illustrated in Fig. 1. The parton model description illustrated in the left part of Fig. 1, is more suitable at low energy where the coherence time of radiation (see below) is short. The results of description of low energy data from fixed target experiments are depicted in Fig. 2. The calculations [1] are parameter free, since the magnitude of broadening is calculated (see details in [1, 2]).

At higher energies, if both x_1 and x_2 are sufficiently small, one can assume that the colliding partons 1 and 2 are gluons. The initial gluon 1 can fluctuate as is depicted in Fig. 1 (right) either as $g \to gg$, or $g \to q\bar{q}$. The latter is a higher order correction within the leading $\log(s)$ dynamics, since misses the $\ln(s)$ enhancement factor, which is contained in the $g_1 \to g_3 g_4$ term due to the integration $d\alpha/\alpha$, where $\alpha = p_+^{(3)}/p_+^{(1)}$ is the fractional light-cone momentum of the produced gluon. Such an α-distribution of radiated gluons results in a small value of the mean fractional momentum α $1/\ln(s)$ 1, neglected in what follows.

Journal of Physics: Conference Series **589** (2015) 012013 doi:10.1088/1742-6596/589/1/012013

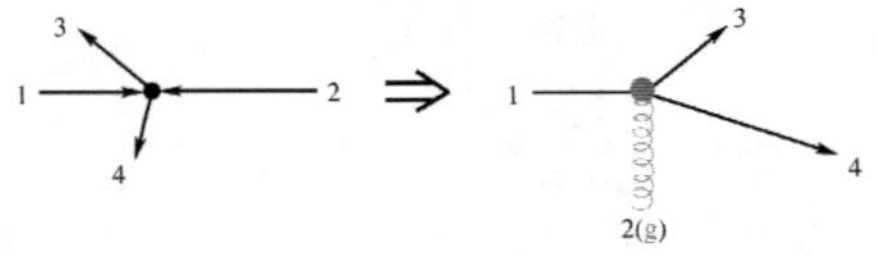

Figure 1. Production of a high-p_T parton in the c.m. frame (left) and in the rest frame of the target (right).

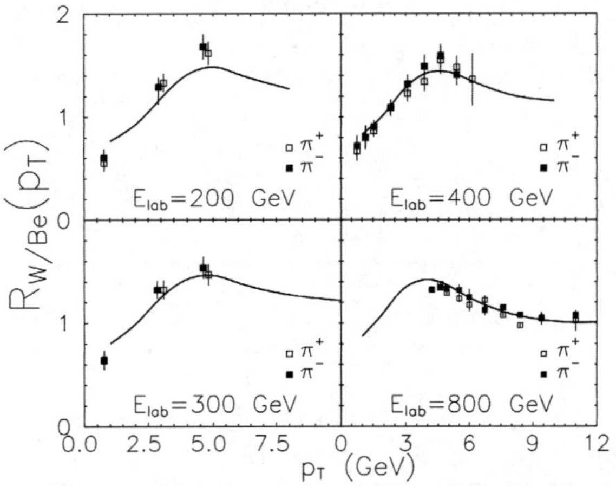

Figure 2. Fractional energy loss by a quark with different initial energies in vacuum vs path length L.

The cross section of parton 3 production integrated over the phase space of parton 4 can be expressed in terms of the interaction cross section of a dipole consisted of partons 3, 4 and $\bar{1}$, $\sigma_{\bar{1}23}(\vec{r}_{13}, \vec{r}_{14})$, where $\vec{r}_{ij}$ is the transverse separation between partons i and j [3]. These distances are kinematically related, $r_{13} = (1-\alpha)r_{34}$, $r_{14} = \alpha r_{34}$.

The cross section of gluon radiation with transverse momentum k_T and rapidity y on nucleon, or nuclear targets can be represented within the dipole approach as, [3, 4, 1],

$$\frac{d\sigma^{N(A)}_{g\to 2g}(\alpha, x_2)}{d^2k_T\, dy} = \int d^2r d^2r'\, e^{i\vec{k}_T(\vec{r}-\vec{r}')} \Big\langle \Psi^\dagger_{gg}(\vec{r}, \alpha)\Psi_{gg}(\vec{r}', \alpha)\Big\rangle\, \Sigma^{N(A)}_{3g}(\vec{r}, \vec{r}', \alpha, x_2) \tag{1}$$

Here $\vec{r} \equiv \vec{r}_{34}$ is the transverse separation between the produced gluons g_3 and g_4; $\vec{r}'$ corresponds to the conjugated amplitude. We rely here on the high-energy approximation of small $\alpha \ll 1$, as was commented above. Correspondingly, the combination $\Sigma_{3g}(\vec{r}, \vec{r}')$ of the dipole cross sections, which depends on the target, a nucleon or a nucleus, has the form [3, 4],

$$\Sigma^{N(A)}_{3g}(\vec{r}, \vec{r}', \alpha) = \sigma^{N(A)}_{3g}(r, \alpha) + \sigma^{N(A)}_{3g}(r', \alpha) - \sigma^{N(A)}_{3g}(\vec{r} - \vec{r}', \alpha); \tag{2}$$

where

$$\sigma^A_{3g}(r, \alpha) = 2\int d^2b \left[1 - e^{-\frac{1}{2}\sigma^N_{3g}(r,\alpha)T_A(b)}\right]. \tag{3}$$

The 3-gluon ($g_1g_2g_3$) dipole cross section is related to the quark-antiquark dipole cross section on a nucleon, known well from phenomenology,

$$\sigma^N_{3g}(r, \alpha) = \frac{9}{8}\Big\{\sigma_{\bar{q}q}(r) + \sigma_{\bar{q}q}(\alpha r) + \sigma_{\bar{q}q}[(1-\alpha)r]\Big\}. \tag{4}$$

All dipole cross sections also depend implicitly on x_2, which is defined later. The impact parameter dependent nuclear thickness function $T_A(b) = \int_{-\infty}^{\infty} dz\, \rho_A(b, z)$ is a result of integration of the nuclear density along the gluon trajectory.

We should remind that we consider here only two Fock components of the physical gluon: (i) just a single gluon, and (ii) two gluons. As a result of interaction, a new final state can be produced (e.g. $g + N \to 2g + X$) only due to the difference between the interactions amplitudes of these two Fock states. The difference has a form of a three-body cross section, this is why

Eqs. (2)-(3) contain the artificial three-gluon dipole cross section, although only one or two gluons interact with the target.

The light-cone distribution function for a 2-gluon fluctuation $g \to 2g$ in Eq. (1) in the limit of small α has the form [4],

$$\Psi_{gg}(\vec{r},\alpha) = \frac{\sqrt{8\alpha_s}}{\pi\, r^2}\exp\left[-\frac{r^2}{2\,r_0^2}\right]\Big[\alpha(\vec{e}_1^{\,*}\!\cdot\vec{e})(\vec{e}_2^{\,*}\!\cdot\vec{r})+(1-\alpha)(\vec{e}_2^{\,*}\!\cdot\vec{e})(\vec{e}_1^{\,*}\!\cdot\vec{r})-\alpha(1-\alpha)(\vec{e}_1^{\,*}\!\cdot\vec{e}_2^{\,*})(\vec{e}\cdot\vec{r})\Big], \quad (5)$$

where $\vec{e}_i$ is the polarization vector of the gluon g_i participating in the process. The parameter $r_0 = 0.3\,\mathrm{fm}$ characterizes the strength of the nonperturbative interaction of gluons. There are many experimental [5] and theoretical [6] evidences that the mean gluon separation is short.

Eventually, the cross section of inclusive hadron production in pp or pA collision can be presented as a convolution of (1) with the gluon distribution function in the projectile proton and with the gluon fragmentation function,

$$\frac{d^2\sigma(pN(A)\to hX)}{d^2p_T\,dy} = \int\limits_{z_{min}}^{1}\frac{dz}{z^2}\,D_{h/g}(z,k_T^2)\int\limits_{x_{min}}^{1}dx_1\,g(x_1,k_T^2)\frac{d^2\sigma^{N(A)}_{g\to 2g}(\alpha,x_2)}{d^2k_T\,dy}. \quad (6)$$

Here $z_{min} = (p_T/\sqrt{s})\,e^y$; $x_{min} = z_{min}/z$; $k_T = p_T/z$. The scale for the gluon distribution and fragmentation function is imposed by gluon transverse momentum k_T. In this integral the fractional momentum of the gluon fragmenting into the detected hadron with rapidity y is related to other variables as,

$$\alpha \equiv \frac{p_+^{(3)}}{p_+^{(1)}} = \frac{p_T\,e^y}{x_1 z\sqrt{s}} = \frac{x_{min}}{x_1}. \quad (7)$$

As far as α is fixed, the value of x_2 is known as well,

$$x_2 = \frac{p_T^2}{x_1 z^2\,\alpha(1-\alpha)s} \approx \frac{p_T}{z\sqrt{s}}\,e^{-y}. \quad (8)$$

1.1. Cronin effect at RHIC: predicted and observed

The ratio of the cross sections Eq. (6) on nuclei to nucleon predicted in [1] for $\sqrt{s} = 200\,\mathrm{GeV}$ at the mid rapidity is plotted in Fig. 3 as function of p_T by dotted curve. The predicted magnitude of the nuclear enhancement was OK, but the shape of p_T-dependence was not.

The employed unintegrated gluon density proposed by Golec-Biernat and Wüsthoff (GBW) [7] peaks at too small p_T. Currently available a more realistic parametrization, for the unintegrated gluon distribution proposed later by Kimber, Martin and Ryskin (KMR) [8, 9] improves the shape. Since no other modifications was made in the computing code, the result still has the status of prediction.

Notice, that [1] was the only successful prediction at the energies of RHIC (e.g. compare with [10]).

1.2. Cronin effect at LHC

The new feature of nuclear effects at the energies of LHC is gluon shadowing, which is a negligibly small correction at RHIC. Gluon shadowing in nuclei is a part of the Gribov inelastic corrections [11] related to the triple-Pomeron term in diffraction. The expected value of the Pomeron-proton cross section is $\sigma_{tot}^{I\!Pp} \sim 50\,\mathrm{mb}$, while measured in single diffraction to large invariant masses turns out to be $\sigma_{tot}^{I\!Pp} < 2\,\mathrm{mb}$! Smallness of the diffractive cross section means weakness of gluon shadowing. In terms of pQCD this shows a suppression of diffractive gluon radiation, which can only be related to smallness of gluonic dipoles ($r_0 = 0.3\,\mathrm{fm}$ in Eq. (5)).

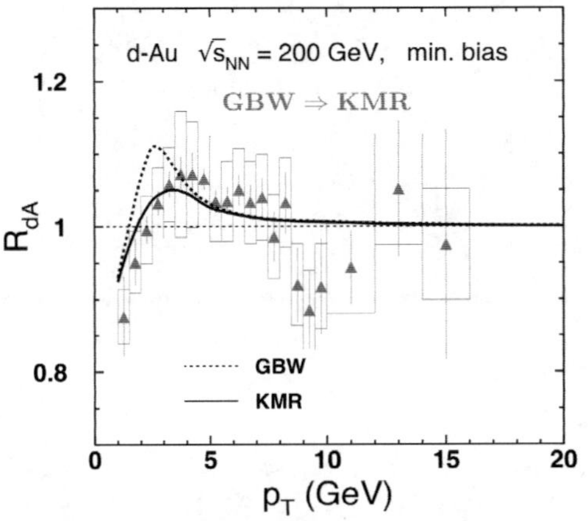

Figure 3. The nuclear to proton ratio R_{dA} as function of p_T. Dashed and solid curves show the predictions made with the GBW [7] and KMR [8, 9] parametrizations for the unintegrated gluon density. Data are from [12].

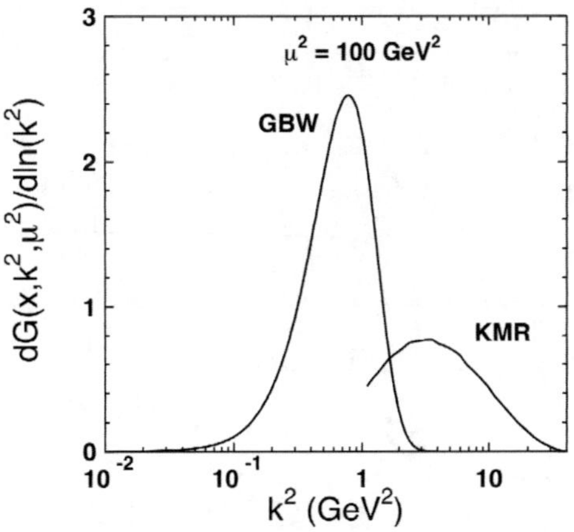

Figure 4. The k^2-dependence of the unintegrated gluon density with the GBW [7] and KMR [8, 9] parametrizations.

Gluon shadowing in DIS corresponds to inclusion of the higher Fock components of the photon, $\gamma^* \to \bar{q}q+g+...$, [4]. Although gluons do not participate in electromagnetic interactions, The gluon PDF is probed by DIS via the DGLAP evolution for the Q^2 dependence of $F_2(x, Q^2)$. So far only the NMC experiment managed to detect a variation of the nuclear PDF with Q^2. However, the data are not sensitive to gluon shadowing (at least in the leading order).

As far as the light-cone distribution function for gluons, Eq. (5), is fixed by diffraction data, the magnitude of gluon shadowing can ve evaluated theoretically [4]. The predictions made in [1] with the cross sections Eq. (6) for $\sqrt{s} = 5\,\text{TeV}$ and the mid rapidity are plotted in Fig. 5 by dotted curve in comparison with data from ALICE experiment [13]. Gluon shadowing, calculated in [4] brings $R_{pA}(p_T)$ down, as is shown by the solid curve.

Similar to the comparison done above for the RHIC data (Fig. 3), the magnitude of the Cronin enhancement is predicted correctly, but the shape of the p_T-dependence does not look correct. Again, the GBW parametrization used in [1] nowadays could be replaced by more realistic KMR form [8, 9] (see Fig. 4). This results in a considerable shift of the maximum to larger p_T, as is shown in Fig. 6. At the same time, the theoretical calculation of gluon shadowing [4] also contains considerable uncertainties. If one applies the nuclear gluon PDF extracted from data in the next-to-leading order analysis by De Florian and Sassot (DS) [14], the results seem to agree with data pretty well, both for shape and magnitude.

All these modifications demonstrate the range of theoretical uncertainties of the prediction made in citecronin-prl for the Cronin effect at the LHC energies. The modified curves still have the status of prediction, because no modification in the calculation was done, except using updated more realistic parametrizations for the phenomenological functions extracted from other reactions.

Notice that the other predictions made recently for the Cronin effect at LHC disagree with the data (see references and comparison with data in [15]).

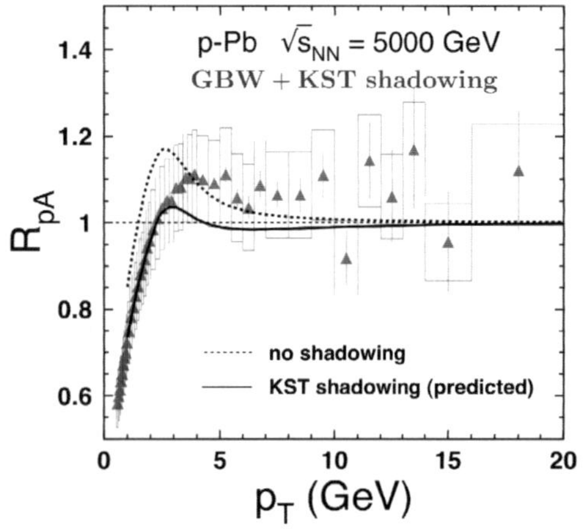

Figure 5. Data for R_{pA} from the ALICE experiment [13] at $\sqrt{s} = 5$ TeV in comparison with predictions made in [1]. Dashed and solid curves show the results calculated without and with gluon shadowing [4].

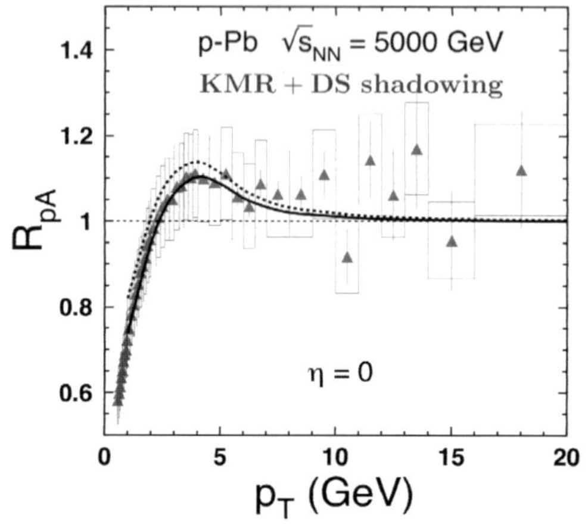

Figure 6. The same as in FIg. 5, but the calulations are performed with the KMR [8, 9] parametrization for the unintegrated gluon density. DS gluon shadowing extracted from data [14].

2. Energy deficit at forward rapidities

Multiple interactions of the projectile hadron and its debris propagating through the nucleus should cause a dissipation of energy. This intuitive expectation is supported by consideration of the Fock state decomposition. The projectile hadron can be expanded over different states which are the fluctuations of this hadron. In the limit of infinite momentum frame those fluctuations live forever. One can probe the Fock state expansion by interaction with a target. The interaction modifies the weights of the Fock states, some interact stronger, some weaker.

In each Fock component the hadron momentum is shared by the constituents, and the momentum distribution depends on their multiplicity: the more constituents are involved, the smaller is the mean energy per a constituent parton, i.e. the softer is the fractional energy distribution of a leading parton. So on a nuclear target the projectile parton distribution falls at $x \to 1$ steeper than on a proton.

In the case of a hard reaction on a nucleus, this softening of the projectile parton momentum distribution can be viewed as an effective loss of energy of the leading parton in the nuclear medium, because the initial state multiple interactions enhance the weight factors for higher Fock states in the projectile hadron. Those components with large number of constituents have a tough energy sharing, so the mean energy of the leading parton decreases compared to lower Fock states, which dominate the hard reaction on a proton target. Such a reduction of the mean energy of the leading parton can be treated as an effective energy loss, which is proportional to the initial hadron energy.

The corresponded suppression factor for each of multiple interactions was evaluated in [16], $S(\xi) \approx 1 - \xi$, where $\xi = \sqrt{x_L^2 + x_T^2}$, and $x_L = 2p_L/\sqrt{s}$; $x_T = 2p_T/\sqrt{s}$. This factor leads to a suppression of the cross section of high-p_T hadron production at forward rapidities. The results of parameter-free calculations performed in [16] depicted in Fig. 7 well agree with data from the BRAHMS [17] and STAR [18] experiments.

Notice that the alternative interpretation [20] of the suppression observed at forward rapidities is based on the effect of coherence (color glass condensate (CGC)). This effect should disappear at lower energies, because $x \propto 1/\sqrt{s}$ rises. Having no other contributing to the suppression

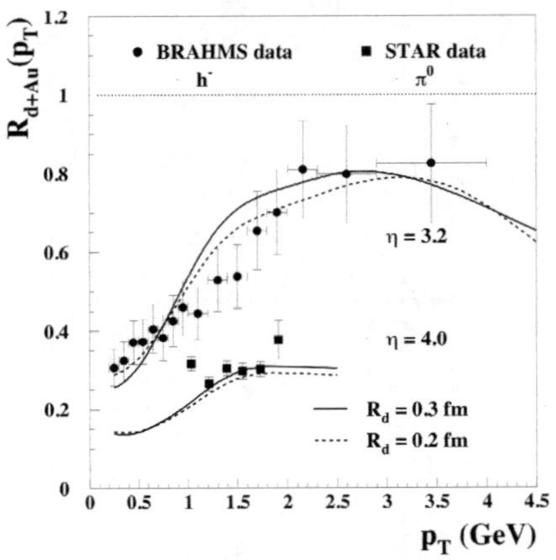

Figure 7. The nuclear ration $R_{dA}(p_T)$ measured for negative charged hadrons produced with pseudo-rapidity $\eta = 3.2$ [17] and for neutral pion produced with $\eta = 4$ [18] in dA collisions at $\sqrt{s} = 200\,\text{GeV}$. The curves show the results of parameter-free calculations done in [16, 19]. Solid and dashed curves correspond to different diquark radiuses.

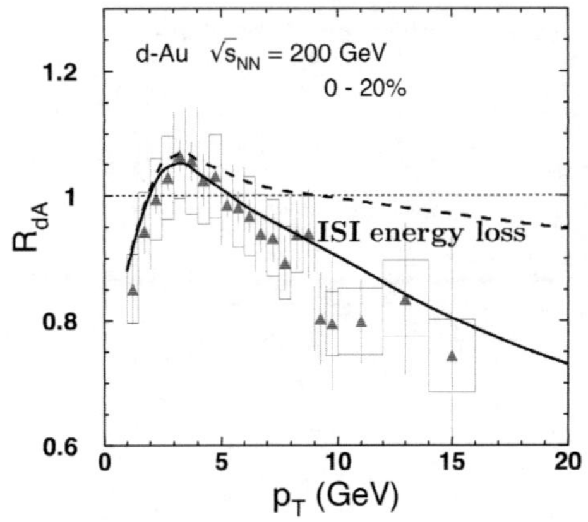

Figure 8. Nuclear ratio $R_{dA}(p_T)$ for neutral pion production in dA central collisions at $\sqrt{s} = 200\,\text{GeV}$ measured in [12]. The dashed curve includes only isotopic effect (deutron vs proton), while the solid curve includes also the initial state energy loss (see the text).

observed by BRAHMS, but only CGC, one should expect no suppression at forward rapidities at lower energies. On the other hand, the suppression caused by energy deficit scales in Feynman x_F and should exist at any energy. Thus lowering the collision energy would be a sensitive test for the models. The NA49 experiment at SPS has performed measurements [21] similar to BRAHMS, but at much lower energy, where the value of x_2 is two orders of magnitude larger than in the BRAHMS data. The results show that the effect of suppression at forward rapidities is still there.

Another test of the mechanisms can be performed at large x_T. In this case no coherence effects are possible, while the value of ξ is considerable and the energy deficit should lead to a suppression at large x_T similar to what was observed at large x_L. The predictions of Ref. [19] at $\sqrt{s} = 200\,\text{GeV}$ and mid rapidity are confirmed by data for central dA collisions with large p_T, as is shown in Fig. 8. No alternative explanation has been proposed so far.

3. High-p_T hadrons from AA collisions

3.1. Hard parton collisions

Parton scattering with a high transverse momentum p_T leads to formation of four cones of gluon radiation: (i) the color field of the colliding partons is shaken off in forward and backward directions; (ii) the scattered partons carry no field up to transverse momenta $k_T < p_T$ ($\vec{k}_T$ is normal to the jet direction $\vec{p}_T$), so the final state partons are regenerating the lost color field by radiating gluons and forming the up-down jets, as is illustrated in Fig. 9.

The coherence length/time of radiation of a gluon carrying fractional light-cone momentum x reads,

$$l_c = \frac{2E\,x(1-x)}{k_T^2 + x^2 m_q^2} \approx \frac{2\,\omega}{k_T^2}, \tag{9}$$

where ω is the gluon energy. First are radiated gluons with small longitudinal and large

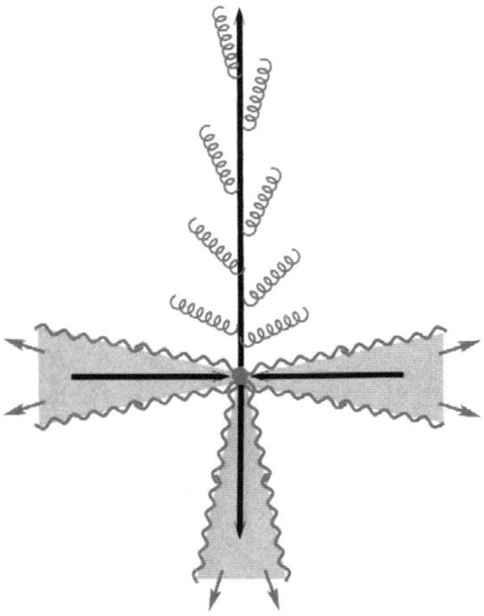

Figure 9. Production of four cones of radiation in a hard parton collision (see the text).

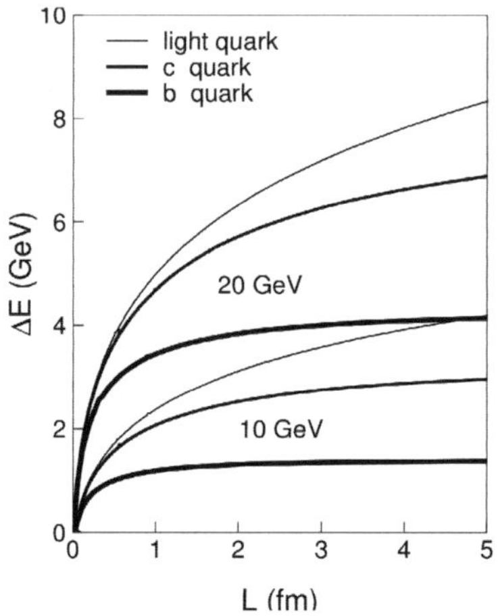

Figure 10. Fractional energy loss by a quark with different initial energies in vacuum vs path length L.

transverse momenta. Energetic gluons with small k_T are radiated at long times.

3.2. Vacuum energy loss

The radiative energy loss is most intensive at the early stage of hadronization, then it slows down. This is confirmed by a direct calculation of the energy radiated over the path length L,

$$\Delta E(L) = E \int\limits_{\Lambda^2}^{Q^2} dk^2 \int\limits_0^1 dx\, x\, \frac{dn_g}{dx\, dk^2} \Theta(L - l_c) \tag{10}$$

where

$$\frac{dn_g}{dx\, dk^2} = \frac{2\alpha_s(k^2)}{3\pi\, x}\, \frac{k^2[1+(1-x)^2]}{[k^2 + x^2 m_q^2]^2} \tag{11}$$

This expression shows that gluon radiation is subject to a dead-cone effect: gluons with $k^2 < x^2 m_q^2$ are suppressed. Heavy quarks radiate less energy than the light ones [22].

Besides, gluon radiation is suppressed by another dead cone effect: soft gluons cannot be radiated at short path length, since the high-p_T parton is lacking its color field, which was shaken off in the forward direction,

$$k^2 > \frac{2Ex(1-x)}{L} - x^2 m_q^2 \,. \tag{12}$$

This is why heavy and light quarks at the early stage radiate with similar rates, up to short length scale [23]

$$L \lesssim \frac{Ex(1-x)}{x^2 m_q^2} \,. \tag{13}$$

Since the rate of energy dissipation is especially large at the early stage, the parton may lose an essential fraction of its initial energy within a short time interval after the hard collision. Indeed the fractional energy loss depicted in Fig. 11 as function of path length shows that a high-p_T quark loses 40% of the total radiated energy during the first 1fm.

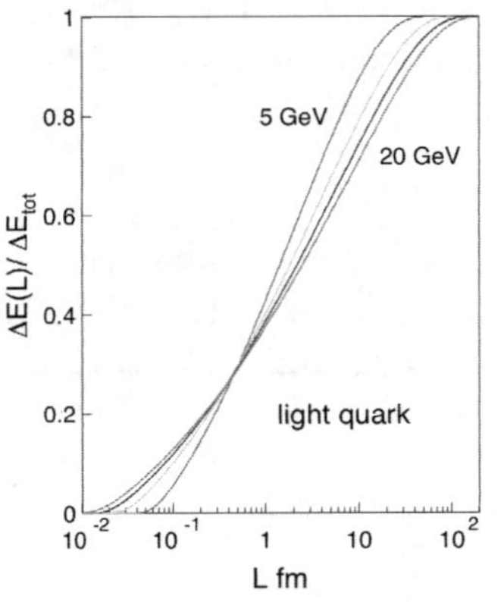

Figure 11. Fractional energy loss by a quark with different initial energies in vacuum vs path length L.

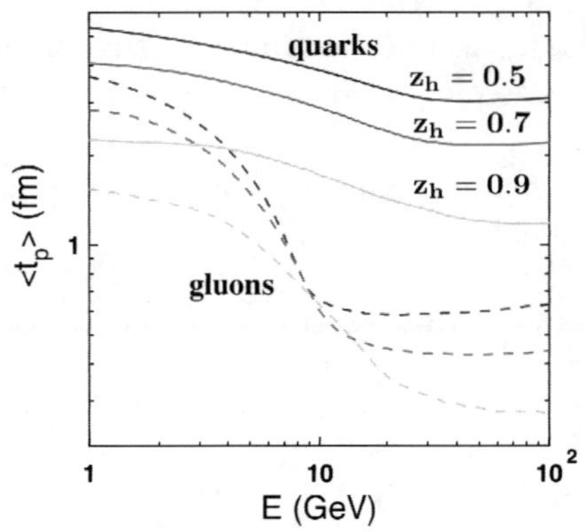

Figure 12. The mean production length vs energy of jets initiated by quarks (solid curves) and gluons (dashed curves). In both cases $z_h = 0.5,\ 0.7,\ 0.9$ (from top to bottom).

This means that Energy conservation restricts the time scale of hadron production. Namely, a hadron with large fractional momentum z_h should be produced at a short time scale, otherwise too much energy will be radiated, and production of the hadron becomes impossible [24, 25, 26]. Notice that in a high-p_T hadron production the large values of z_h are favored by the steeply falling transverse momentum spectrum of quarks convoluted with the fragmentations function [26, 27, 28].

In fact the energy restrictions make the dissipation of energy for a given z_h somewhat slower than it follows from Eq. (10) and is depicted in Figs. 10, because radiation of gluons with fractional momenta larger than $1 - z_h$ is forbidden by energy conservation [24, 29]. On the other hand, a ban for radiation of gluons with energy $\omega > (1 - z_h)E$ in (10), leads to the Sudakov type suppression factor [25, 29],

$$S(L, z_h) = \exp\left[-\langle n_g(L, z_h)\rangle\right] , \tag{14}$$

where $\langle n_g(L, z_h)\rangle$ is the mean number of nonradiated gluons during propagation over the distance L,

$$\langle n_g(L, z_h)\rangle = \int\limits_{1/Q}^{l_{max}} dl \int\limits_{(2El)^{-1}}^{1} d\alpha\, \frac{dn_g}{dl d\alpha}\, \Theta\left(\alpha + \frac{1-\alpha}{2lE} - 1 + z_h\right) . \tag{15}$$

Here α is the fractional light-cone momentum of a radiated gluon; $l_{max} = \min\{L,\ E/2\lambda^2\}$, and λ is the soft cutoff for transverse momenta of gluons, fixed in [29] at $\lambda = 0.7\,\mathrm{GeV}$. The step function in (15) takes care of energy conservation.

Combination of these two effects, vacuum energy loss and Sudakov suppression, leads to a rather short production length, which slightly varies with jet energy and virtuality. The resulting production length distribution for quark jets with maximal virtuality $Q = p_T/z_h = E$ is depicted in Fig. 12 as function of jet energy [29, 27, 28]. Several examples for quark jets are depicted by solid curves. The mean production length for gluon jets should be shorter, because the more intensive vacuum energy loss and a stronger Sudakov suppression act in the same direction, making the $\langle l_p\rangle$ shorter. The results for gluon jets are shown in Fig. 12 by dashed curves. We see that the mean production length is rather short, and is slowly decreasing with energy (i.e. with the hadron p_T). Therefore, a colorless dipole, which does not lose energy anymore and

Journal of Physics: Conference Series **589** (2015) 012013 doi:10.1088/1742-6596/589/1/012013

later develops the wave function of the detected hadron, is produced at the early stage of jet development and attenuates in the medium. This, rather than induced energy loss, may be the main source of the observed suppression of high-p_T hadrons produced in AA collisions.

3.3. Attenuation of high-p_T hadrons in a hot medium

The transverse size of a colorless dipole fluctuates during its propagation through a medium, and the attenuation rate is varying. The way to describe this process is the path-integral method, summing up all possible trajectories of the dipole constituents [30, 3, 4]. The imaginary part of the light-cone potential, responsible for absorption, is controlled by the transport coefficient $\hat{q}$, which is defined as the rate of transverse momentum broadening of a parton propagating through the medium. The dependence of $\hat{q}$ on transverse coordinate and time passed after the hard collision is usually modeled as [31],

$$\hat{q}(l,\vec{b},\vec{\tau}) = \frac{\hat{q}_0\, l_0}{l}\,\frac{n_{part}(\vec{b},\vec{\tau})}{n_{part}(0,0)}\,\Theta(l-l_0), \tag{16}$$

where $\vec{b}$ is the impact parameter of nuclear collision, $\vec{\tau}$ is the impact parameter of the hard parton-parton collision relative to the center of one of the nuclei, $n_{part}(\vec{b},\vec{\tau})$ is the number of participants, and $\hat{q}_0$ is the rate of broadening of a quark propagating in the maximal medium density produced at impact parameter $\tau = 0$ in central collisions ($b = 0$) at the time $t = t_0 = l_0$ after the collision. The corresponding transport coefficient for gluons should be 9/4 bigger. The equilibration time t_0 is model dependent. The results are not very sensitive to it, so it is fixed it at $t_0 = l_0 = 0.5\,\mathrm{fm}$.

$$R_{AB}(\vec{b},p_T) = \frac{\int d^2\tau\, T_A(\tau) T_B(\vec{b}-\vec{\tau}) \int\limits_0^{2\pi} \frac{d\phi}{2\pi} \left| \int\limits_0^1 d\alpha \int d^2r_1 d^2r_2\, \Psi_h^\dagger(\vec{r}_2,\alpha) G_{\bar{q}q}(l_1,\vec{r}_1;l_2,\vec{r}_2)\Psi_{in}(\vec{r}_1,\alpha)\right|^2}{T_{AB}(b) \left| \int\limits_0^1 d\alpha \int d^2r\, \Psi_h^\dagger(\vec{r}_2,\alpha)\Psi_{in}(\vec{r}_1,\alpha)\right|^2}\,. \tag{17}$$

Here $G_{\bar{q}q}(l_1,\vec{r}_1;l_2,\vec{r}_2)$ is the Green function describing proparagion of a $\bar{q}q$ dipole between the longitudinal coordinates (in the medium rest frame) l_1 and l_2 having the initial and final transverse separations $\vec{r}_{1,2}$; Ψ_{in} and Ψ_h are the light-cone distribution functions of the initial and final (the hadron) dipoles; α is the fractional light-cone momentum; $T_{AB} = \int d^2\tau\, T_A(b)T_B(\vec{b}-\vec{\tau})$; ϕ is the azimuthal angle of the dipole trajectory in impact parameter plane, relative to the impact vector $\vec{b}$ of the collision. The imaginary part of the light-cone potential controlling the solution for the Green function is proportional to the transport coefficient [27, 28] $\hat{q}(l,\vec{b},\vec{\tau})$ along the trajectory, Eq. (16).

The results of calulation [28] at $\sqrt{s} = 2.76\,\mathrm{TeV}$ with $\hat{q}_0 = 2\,\mathrm{GeV}^2/\mathrm{fm}$ are plotted in Fig. 13 by dashed curve in comparison with data from the ALICE [32] and CMS [33, 34] experiments. While $\hat{q}_0$ controls the normalization of $R_{AA}(p_T)$, the shape of the p_T-dependence is parameter free and well reproduces the data.

With the fixed value of the parameter $\hat{q}_0$ one also reproduces well the p_T-dependence of R_{AA} observed at different centralities of collision (see detailes in [28]).

Similar calculations performed at $\sqrt{s} = 200\,\mathrm{GeV}$ with $\hat{q}_0 = 1.6\,\mathrm{GeV}^2/\mathrm{fm}$ are depicted in Fig. 15 by dashed curve. Data do not support such a steep rise of R_{AA} with p_T.

3.4. Azimuthal asymmetry

As far as the produced hadron suppression correlates with the path of the dipole in the hot absorptive medium, this leads to an azimuthal asymmetry for hadron production from the

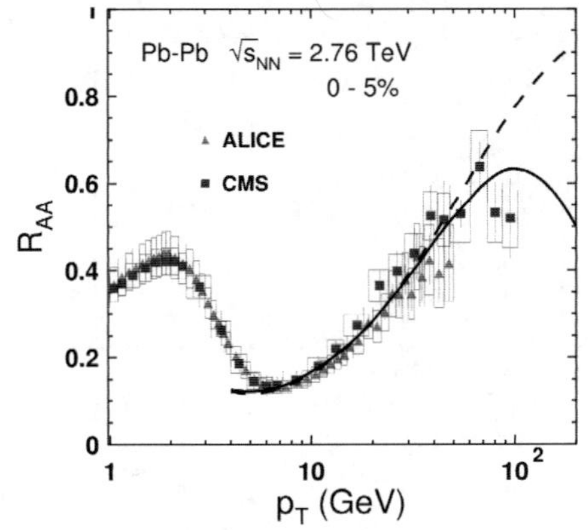

Figure 13. The suppression factor R_{AA} for central lead-lead collisions at $\sqrt{s} = 2.76\,\mathrm{TeV}$. The dashed line is calculated within the path-integral approach [28] with the transport coefficient Eq. (16), where the adjusted parameter $\hat{q}_0 = 2\,\mathrm{GeV}^2/\mathrm{fm}$. The solid curve also includes the effects of initial state interactions in nuclear collisions [16, 19]. Data are from the ALICE [32] and CMS [33, 34] experiments.

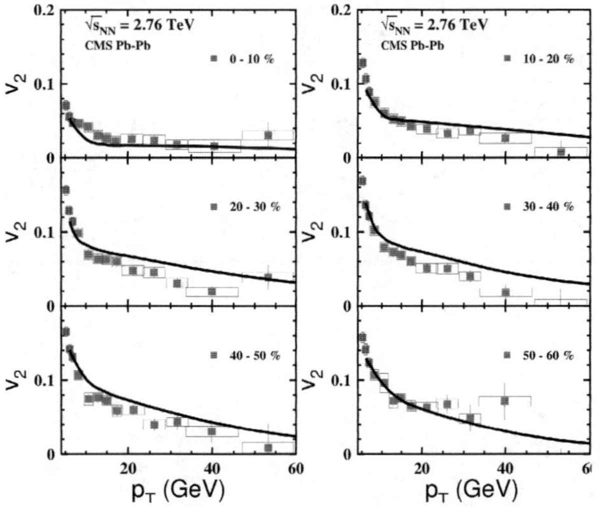

Figure 14. CMS data [35] for azimuthal anisotropy, v_2, vs p_T for charge hadron production in lead-lead collisions at mid rapidity, at $\sqrt{s} = 2.76$ TeV and at different centralities indicated in the figure. The curves present the results of calculation with Eq. (18) and $\hat{q}_0 = 2\,\mathrm{GeV}^2/\mathrm{fm}$ [28].

almond-shaped overlap in a non-central collision of nuclei. It can be calculated similarly to Eq. (17),

$$v_2(p_T, b) = \frac{\int d^2\tau\, T_A(\tau) T_B(\vec{b} - \vec{\tau}) \int\limits_0^{2\pi} d\phi \cos(2\phi) \left| \int\limits_0^{\infty} dr\, r\, \Psi_h(r) G_{\bar{q}q}(0,0; l_{max}, r) \right|^2}{\int d^2\tau\, T_A(\tau) T_B(\vec{b} - \vec{\tau}) \int\limits_0^{2\pi} d\phi \left| \int\limits_0^{\infty} dr\, r\, \Psi_h(r)\, G_{\bar{q}q}(0,0; l_{max}, r) \right|^2}. \tag{18}$$

Comparison with the data, performing another sensitive test of the model, is presented in FIg. 14. The agreement at large p_T is rather good.

3.5. Energy deficit at high p_T

Similar to the effects of energy deficit observed at large p_T in dA collisions at RHIC (Fig. 8), production of hadrons at large x_T should also be suppressed. At $\sqrt{s} = 5\,\mathrm{TeV}$ such an additional suppression becomes significant at rather large $p_T \gtrsim 100\,\mathrm{GeV}$, as is shown by solid curve in Fig. 13. The points at largest measured p_T indeed seem to deviate from the dashed curve.

Naturally, the lower is the collision energy, the earlier onsets the effect of energy deficit, because it scales in x_T. The results at $\sqrt{s} = 200\,\mathrm{GeV}$ including the energy loss corrections are plotted Fig. 15 by solid curve in comparison with data. Apparently, the additional suppression caused by energy deficit improves agreement with the data.

Even a stronger effect of energy loss is expected at lower collision energies. Our predictions at $\sqrt{s} = 62\,\mathrm{GeV}$ with $\hat{q}_0 = 1.2\,\mathrm{GeV}^2/\mathrm{fm}$ corrected for energy deficit are compared with data at difference centralities in Fig. 16. We see that the energy loss corrections are so strong that cause a falling, rather than rising, p_T-dependence of R_{AA}.

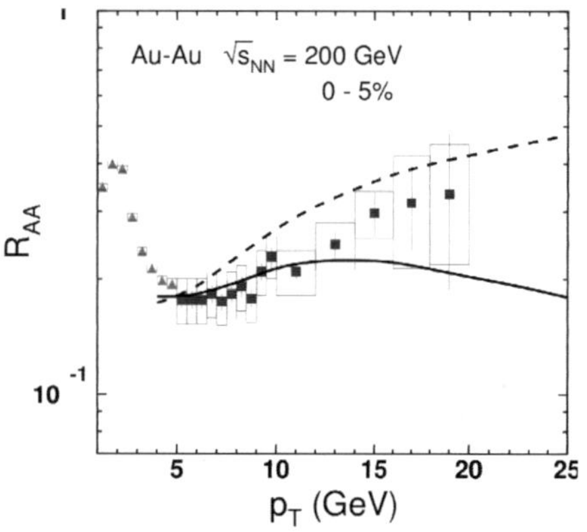

Figure 15. Nuclear attenuation factor $R_{AA}(p_T)$ for neutral pions produced in central gold-gold collisions at $\sqrt{s} = 200$ GeV. The solid and dashed line are calculated with or without ISI corrections. PHENIX data are from [36] (triangles)

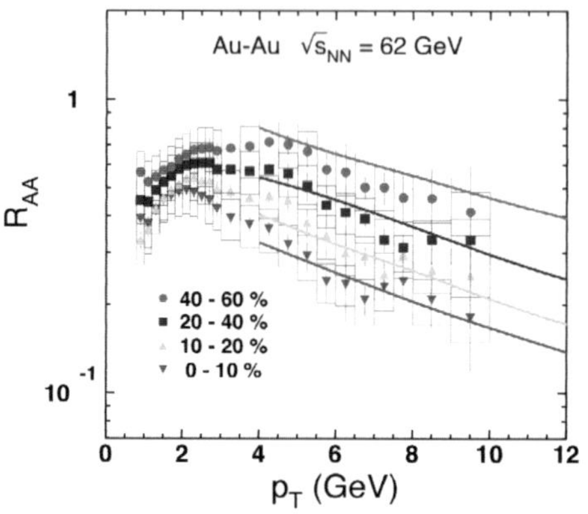

Figure 16. Centrality dependence of the suppression factor $R_{AA}(p_T, b)$ in gold-gold collisions at $\sqrt{s} = 62.4\,\text{GeV}$ [36]. The curves are calculated including FSI attenuation of dipoles in the dense medium and the ISI energy loss effects.

4. Summary

Summarizing, in this note we investigated the impact of energy conservation in several high-energy processes involving nuclei. In many instances we observed significant effects:

- The color-dipole description of high-p_T hadron production is tested on the Cronin effect at the fixed target and collider energies. The description is found to be quite successful and having a strong predictive power. In particular, the magnitude of the Cronin enhancement was successfully predicted prior the measurements at RHIC and LHC.
- Energy conservation significantly suppresses cross sections upon approaching the kinematic bounds, at large x_L and/or at large x_T. The presented parameter-free calculations well explain the observed suppression seen in data from RHIC for hadron production at forward rapidities, or at the mid-rapidity, but very large p_T. Predictions for the energy of LHC are presented as well.
- Inclusive production of high-p_T hadrons in heavy ion collision is subject to final state interactions with the dense quark-gluon medium created in the collision. The popular energy loss interpretation, based on the unjustified assumption of a long hadronization length, conflicts with energy conservation because of the intensive energy dissipation by a highly virtual parton produced in a high-p_T process. The inclusively detected hadron, having a large mean fractional momentum $\langle z_h \rangle$, cannot be produced on a long time scale and respect energy conservation. As far as the production of a colorless dipole occurs on a short length scale, the dipole attenuation and color transparency become the driving dynamics of the observed hadron suppression. The path-integral technique for calculation of the dipole attenuation results in a successful description of available data for nuclear quenching of high$-p_T$ hadrons produced in AA collisions at the energies of RHIC and LHC. The parameter, the transport coefficient $\hat{q}$, characterizing the interaction with the medium, is found at a reasonable magnitude, expected in the theory, while the energy loss scenario has been always suffering of a gross over-evaluation of this parameter. The observed azimuthal asymmetry at RHIC and LHC is well explained as well, without any further adjustment.
- Energy conservation and energy sharing restrictions produce an additional suppression in the case of heavy ion collisions as well. These effects are found especially strong at the

Journal of Physics: Conference Series **589** (2015) 012013 doi:10.1088/1742-6596/589/1/012013

energies of RHIC $\sqrt{s} \leq 200\,\text{GeV}$, but also are expected at LHC, either beyond the currently measured range of p_T, or at forward rapidities.

Acknowledgments

This work was supported in part by Fondecyt (Chile) grants 1130543, 1130549, 1100287, and by Conicyt-DFG grant No. RE 3513/1-1. The work of J.N. was partially supported by the grant 13-20841S of the Czech Science Foundation (GAČR), by the Grant MSMT LG13031, by the Slovak Research and Development Agency APVV-0050-11 and by the Slovak Funding Agency, Grant 2/0020/14.

References

[1] Kopeliovich B Z, Nemchik J, Schäfer A and Tarasov A V 2002 *Phys. Rev. Lett.* **88** 232303
[2] Johnson M B, Kopeliovich B Z and Tarasov A V 2001 *Phys. Rev.* C **63** 035203
[3] Kopeliovich B Z, Tarasov A V and Schäfer A 1999 *Phys. Rev.* C **59** 1609
[4] Kopeliovich B Z, Schäfer A and Tarasov A V 2000 *Phys. Rev.* D **62** 054022
[5] Kopeliovich B Z, Potashnikova I K, Povh B and Schmidt I 2007 *Phys. Rev.* D **76** 094020
[6] Di Giacomo A and Panagopoulos H 1992 *Phys. Lett.* B **285** 133
[7] Golec-Biernat K J and Wüsthoff M 1998, *Phys. Rev.* D **59** 014017
[8] Kimber M A , Martin A D and Ryskin M G 2001 *Phys. Rev.* D **63** 114027
[9] Martin A D, Ryskin M G and Watt G 2010 *Eur. Phys. J.* C **66** 163
[10] Kharzeev D, Levin E and McLerran L 2003 *Phys. Lett.* B **561** 93
[11] Gribov V N 1969 *Sov. Phys. JETP* **29** 483 [*Zh. Eksp. Teor. Fiz.* **56** 892].
[12] Adler S S *et al* [PHENIX Collaboration] 2007 *Phys. Rev. Lett.* **98** 172302
[13] Abelev B *et al.* [ALICE Collaboration] 2013 *Phys. Rev. Lett.* **110**, 082302
[14] de Florian D and Sassot R 2004 *Phys. Rev.* D **69** 074028
[15] Albacete J L *et al.* 2013 *Int. J. Mod. Phys.* E **22** 1330007
[16] Kopeliovich B Z, Nemchik J, Potashnikova I K, Johnson M B and Schmidt I 2005 *Nucl. Phys.* B **146** 171
[17] Arsene I *et al* [BRAHMS Collaboration] 2004 *Phys. Rev. Lett.* **93** 242303
[18] Adams J *et al* [STAR Collaboration] 2006 *Phys. Rev. Lett.* **97** 152302
[19] Kopeliovich B Z, Nemchik J 2011 *J. Phys. G* **38** 043101
[20] Kharzeev D, Kovchegov Y V and Tuchin K 2004 *Phys. Lett.* B **599** 23
[21] Boimska B [NA49 Collaboration] 2004 *Ph.D. Dissertation* **CERN-THESIS-2004-035**
[22] Dokshitzer Y L and Kharzeev D E 2001 *Phys. Lett.* B **519** 199
[23] Kopeliovich B Z, Potashnikova I K and Schmidt I 2007 *Phys. Rev.* C **82** 037901
[24] Kopeliovich B Z, Nemchik J and Predazzi E 1996 *Proceedings of the workshop on Future Physics at HERA* ed. by Ingelman G, De Roeck A and Klanner R *DESY 1995/1996* v.2 1038
[25] Kopeliovich B Z, Nemchik J, Predazzi E and Hayashigaki A 2004 *Nucl. Phys.* A **740** 211
[26] Kopeliovich B Z, Nemchik J and Schmidt I 2007 *Nucl. Phys.* A **782** 224
[27] Kopeliovich B Z, Potashnikova I K and Schmidt I 2011 *Phys. Rev.* C **83**, 037901
[28] Kopeliovich B Z, Nemchik J, Potashnikova I K and Schmidt I 2012 *Phys. Rev.* C **86** 054904
[29] Kopeliovich B Z, Pirner H J, Potashnikova I K and Schmidt I 2008 *Phys. Lett.* B **662** 117
[30] Kopeliovich B Z and Zakharov B G 1991 *Phys. Rev.* D **44** 3466
[31] Chen X F, Greiner C, Wang E, Wang X N and Xu Z 2010 *Phys. Rev.* C **81** 064908
[32] Abelev B *et al.* [The ALICE Collaboration] 2013 *Phys. Lett.* B **720** 52
[33] Lee Y J (for the CMS Collaboration) 2011 *J. Phys. G* **38** 124015
[34] A. S. Yoon A S (for the CMS Collaboration) 2011 *J. Phys. G* **38** 124116
[35] Chatrchyan S *et al.* [CMS Collaboration] 2012 *Phys. Rev. Lett.* **109** 022301
[36] Adare A *et al.* [PHENIX Collaboration] 2013 *Phys. Rev.* C **87** 034911
[37] M. L. Purschke M L (for the PHENIX Collaboration) 2011 *J. Phys. G* **38** 124016
[38] PHENIX Collaboration 2012 preliminary data posted at www.phenix.bnl.gov/WWW/plots/show_plot.php?editkey=p1118

Journal of Physics: Conference Series **589** (2015) 012014 doi:10.1088/1742-6596/589/1/012014

Upgrade and physics perspective of ALICE at the LHC

Levente Molnar for the ALICE Collaboration
23 rue du Loess - BP28, 67037 Strasbourg Cedex 2, France
E-mail: levente.molnar@iphc.cnrs.fr

Abstract. ALICE at the LHC, collecting data in Pb-Pb, p-Pb and pp collisions, aims for the characterisation of the strongly interacting matter at unprecedented energy densities. While the first running period until 2018 can provide detailed description of the global and bulk phenomena and a first set of intriguing results on rare probes of heavy-ion collisions, many important questions involving rare processes cannot be addressed in details with the current experimental setup. In this talk, we give an overview of the ALICE Upgrade program focusing on the physics prospects related to heavy-quarks, quarkonia, low-mass dileptons, jets and exotica searches achievable in high luminosity Pb-Pb collisions at the LHC. We also discuss the technological challenges and choices of the detector upgrades to be installed during the LHC Long Shutdown 2.

1. Introduction

ALICE has delivered new exciting experimental results from high-energy heavy-ion collisions at the Large Hadron Collider (LHC). The ALICE operation in the LHC Run 1 (2009-2013) focused on the light quark sector, such as establishing the presence of strong elliptic flow [1, 2] and strong quenching of high momentum charged particles at LHC [3], exceeding the values measured at RHIC. Moreover, the first measurements in the heavy quark sector brought intriguing results, such as the observed D meson elliptic flow, and quenching of identified open charm mesons [4, 5] and the quarkonia suppression at forward rapidities with a different trend than shown by the RHIC results [6].

The ALICE data taking in the LHC Run 2 (2015-2018) before the LHC Long Shutdown 2 (LS2), most likely will bring comprehensive and completed results on the experimental probes of the prospected ALICE baseline physics program: anisotropic flow of light hadrons, inclusive momentum spectra and nuclear modification of the heavy flavour mesons (D mesons), global jet properties in heavy-ion collisions. While the first characterisation of the heavy quark sector could be achieved with the expected statistics of 1 nb^{-1} delivered integrated luminosity by the LS2 precision measurements would require higher statistics $\sim$10 nb^{-1} in Pb-Pb collisions at top LHC energy.

ALICE is actively pursuing its upgrade program for LS2 and beyond, to perform precision measurements of the following physics topics in heavy-ion collisions: study of heavy quark thermalisation, measurement of low mass dileptons and thermal photons, measurement of gamma-jets with particle identification and search for exotic heavy nuclear states. Detailed description of the general ALICE Upgrade program and the physics goals can be found in [7] and more specific topics related to the upgraded Inner Tracking System in [8, 9].

Journal of Physics: Conference Series **589** (2015) 012014 doi:10.1088/1742-6596/589/1/012014

2. Upgrade strategy

In the currently ongoing LHC Long Shutdown 1 (LS1, 2013-2014) ALICE completes its detector setup, by finishing the installation of the Transition Radiation Detector modules (TRD), adding the last module of the Photon Spectrometer (PHOS) and installing new electromagnetic calorimeter modules (DCAL) opposite the existing ones (EMCAL) [10], allowing improved di-jet and gamma-jet measurements.

The ALICE detector upgrade will take place in the LS2 (2018-2019). The upgrade will enhance the rate capability of ALICE to be able to record and inspect Pb-Pb collisions at a high rate of 50 kHz, since the interesting physics observables are mostly signals at low transverse momentum with a large background, and thus untriggerable (such as the charmed baryons or low mass dileptons). To collect reference data for heavy-ion measurements, the rate capabilities are extended up to few 100 kHz for high luminosity pp and p-Pb collisions.

The ALICE Upgrade consists of new detectors and changes to the existing ones [7]. To allow the installation of a new Inner Tracking System (ITS), the current beam pipe will be replaced with a smaller diameter one. The current ITS will be fully replaced with a new upgraded detector. The Time Projection Chamber (TPC) read-out chambers will be replaced with Gas Electron Multiplier (GEM) read-out chambers to cope with the required faster read-out rates. A new forward tracking system towards the Muon Spectrometer will be installed: the Muon Forward Tracker (MFT). The trigger detectors (V0, T0) will also undergo a general upgrade. The remaining detectors will upgrade their electronics to conform with the increased rate requirements [7].

The online-offline (O^2) systems will also undergo a structural change, merging the data acquisition, the trigger system and the offline system functionalities into a unified framework. The data acquisition will handle continuous and triggered detector read-out in combination with a new central trigger system and a new high level trigger infrastructure performing on-the-fly reconstruction. The O^2 system has to handle a factor of 100 more statistics in the recorded data then present.

3. Detector projects

3.1. New Inner Tracking System

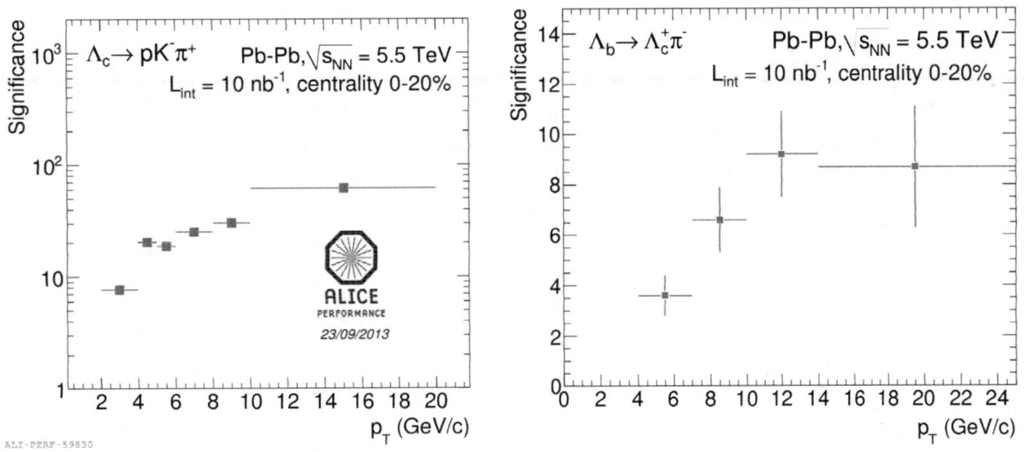

Figure 1. Expected significance of the $\Lambda_c \to pK^-\pi^+$ (left panel) and $\Lambda_b \to \Lambda_c^+\pi^-$ measurements as a function of p_T in the 0-20% centrality for Pb-Pb collisions at $\sqrt{s_{NN}} = 5.5$ TeV [9].

The new beam-pipe will allow the installation of an upgraded ITS, closer to the interaction point compared to the current detector setup (present first layer is at 39 mm radial position with respect to the upgraded one at 22 mm). The upgraded ITS will consist of seven layers with a largely reduced material budget with respected to the current ALICE-ITS and the upgraded inner tracker setup of ATLAS and CMS [8, 9].

The reduction in the material budget is achievable using Monolithic Active Pixels Sensors (MAPS) with a silicon material budget of 50 μm instead of the current 350 μm. Furthermore, optimisation of the electrical power and signal cable layouts, the mechanical and cooling structures leads to the 0.3 X/X_0 target material budget for the 3 inner layers and 0.8 X/X_0 for the 4 outer layers. The layers will consists of high granularity pixel sensors with pixel size from 20 μm x 20 μm up to 50 μm x 50 μm, operated in continuous read-out mode.

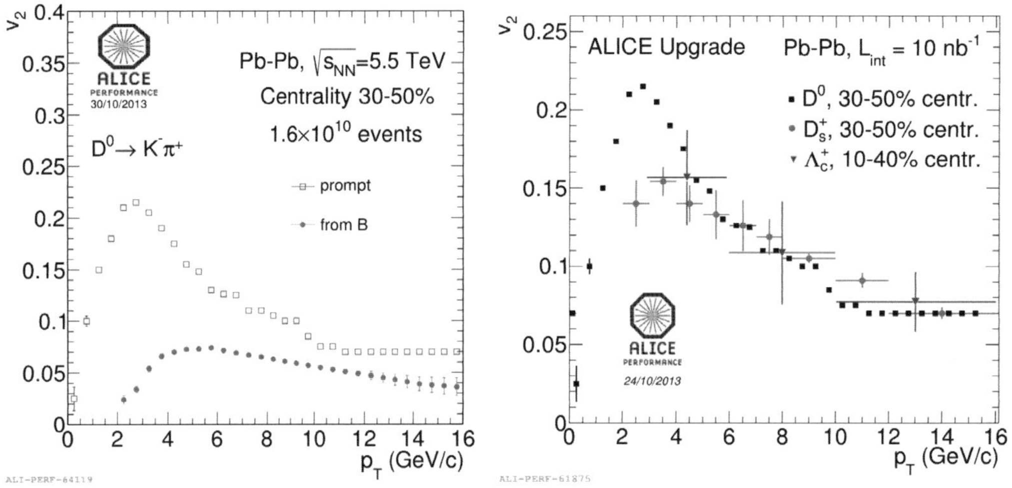

Figure 2. Comparison of the estimated statistical uncertainties on elliptic flow of prompt and secondary D mesons (left panel) and of prompt D mesons with D_s^+ and Λ_c^+ (right panel) in the 30-50% centrality in Pb-Pb collisions at $\sqrt{s_{NN}} = 5.5$ TeV [8, 9].

The new detector design, with the required 4-6 μm spatial resolution, will allow a factor of 3 (in rϕ) and 5 (in z) improvement in the vertex resolution, a factor of 3 improvement in the standalone tracking momentum resolution and a high ~83 % standalone tracking efficiency at $p_T = 100$ MeV/c. These improvements give access to the measurement of charmed and beauty baryons (typical life-time of O(10) μm) and to the discrimination of prompt and secondary heavy quark production. Fig. 1 shows an example on the expected significance of the Λ_c^+ and Λ_b measurements, that will be accessible in Pb-Pb collisions with the ALICE Upgrade and high rate data taking at top LHC energy. Fig. 2 shows the expected separation and statistical uncertainties of prompt and secondary D meson elliptic flow measurements and of prompt D meson, D_s^+ and Λ_c^+ elliptic flow measurements with the upgraded ITS at top LHC energy in Pb-Pb collisions. The detailed physics program and performance of the upgraded ITS is described in [9].

3.2. New GEM read-out for the TPC

The Time Projection Chamber in ALICE [10] is a unique tracking and particle identification detector among the LHC experiments, providing high precision tracking and energy loss measurements.

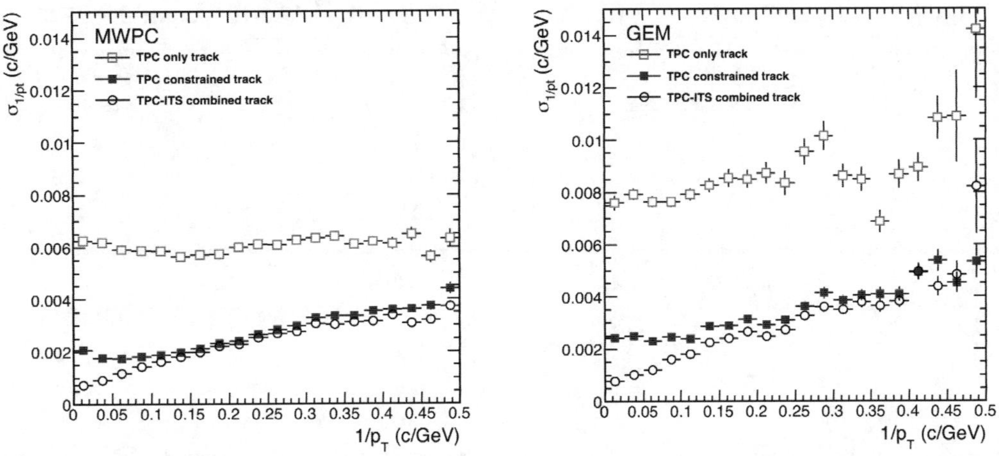

Figure 3. Comparison of the momentum resolution as a function of $1/p_T$ for the present MWPC (left panel) and the proposed GEM (right panel) read-out chambers. The red squares represent track information only from the TPC, the blue squares represent track information only from the TPC including the vertex point in the track fit, and the black circles represent tracks combining the ITS and TPC information [7].

The existing read-out chambers are multi-wire proportional read-out chambers (MWPC) operated in a gated mode to separate the charge collection of electrons and ions [10]. The gated mode is required, otherwise the large space charge build-up, due to ion back flow, would cause field distortions that would lead to deteriorated tracking performance. The gated operation and the current gas mixture employed in the drift volume limit the TPC read-out rate to a few kHz.

To maintain the tracking and particle identification capabilities in high rate operations, the read-out MWPCs and the read-out electronics will be replaced. The MWPCs will be substituted by triple Gas Electron Multiplier (GEM) read-out chambers, operated in a continuous, triggerless read-out mode. GEMs would allow a TPC operation at 50 kHz in Pb-Pb collisions due to their intrinsic ion-feedback suppression [7]. The expected momentum resolution of the GEM setup is shown in Fig. 3 in comparison with the present MWPC setup. The tracking performance is similar for tracks reconstructed from the combined ITS and TPC space points, noting that the current pad plane layout was designed for the MWPC read-out chambers. Its optimisation for the GEM read-out chambers is expected to further improve the performance.

3.3. New forward tracking for the Muon Spectrometer

To improve the physics capabilities of the Muon Spectrometer of ALICE [10], located at large rapidities (-4.0 $< \eta <$ -2.5), a new detector is proposed: the Muon Forward Tracker (MFT) [11], adding secondary vertex reconstruction capability for muon tracks. The MFT will be located upstream to the hadron absorber of the Muon Spectrometer, close to the upgraded ITS. The MFT will consist of five tracking planes, built on the same MAPS technology as the upgraded ITS with similar material budget per plane: 0.4 X/X_0 and spatial resolution <5 μm.

The Muon Spectrometer with MFT will perform quarkonia (J/ψ, ψ') and differential low mass dimuon measurements down to very low transverse momenta, complementary in rapidity range with the upgraded ITS and complementary both in rapidity and in transverse momenta with other LHC experiments. The addition of MFT to the Muon Spectrometer is expected to improve the signal-over-background (S/B) and systematic errors for quarkonia measurements,

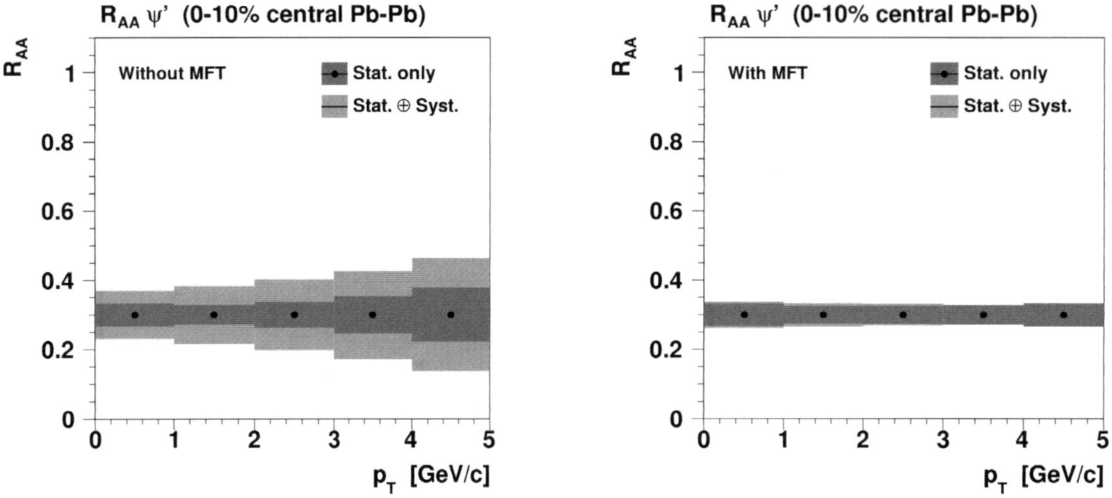

Figure 4. Expected statistical and systematical uncertainties of the nuclear modification factor R_{AA} of the ψ' in central Pb-Pb collisions at $\sqrt{s_{NN}} = 5.5$ TeV without MFT (left panel) and with MFT (right panel) for a 10 nb^{-1} integrated luminosity [11, 12].

eg. in case of ψ' by a factor of 3 - 10 depending on the transverse momentum. Fig. 4 shows the expected reduction of statistical and systematic uncertainties of the ψ' nuclear modification factor in central Pb-Pb collisions at top LHC energy. The low mass dimuon measurements also benefits from the MFT capabilities with the improved combinatorial background suppression, originating from semi-muonic decays of pions and kaons, yielding an increased S/B by a factor of 10 and improved mass resolution by a factor of 3-5 [11, 12]. The improvement of the mass resolution due to MFT is shown in Fig. 5.

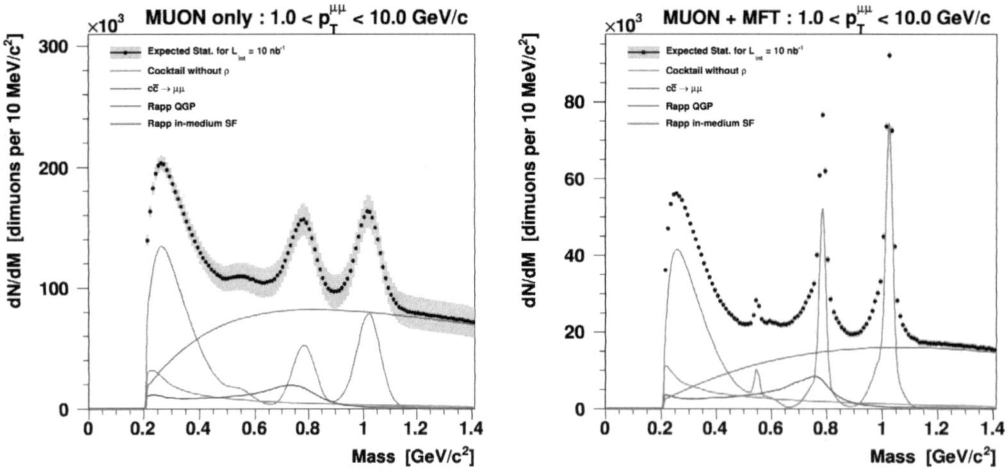

Figure 5. Expected low mass dimuon spectrum (combinatorial background is subtracted) in 0-10% most central Pb-Pb collisions at $\sqrt{s_{NN}} = 5.5$ TeV, normalised to 10 nb^{-1} integrated luminosity without MFT (left panel) and with MFT (right panel) [11, 12].

Journal of Physics: Conference Series **589** (2015) 012014 doi:10.1088/1742-6596/589/1/012014

3.4. Prospects for forward physics at ALICE

While ALICE is actively pursuing the upgrade goals for the LS2, new studies are carried out on a Forward Calorimeter (FoCal) to add direct photon and neutral pion measurement capabilities in the forward region $\sim 3 < \eta < 5$, opposite to the Muon Spectrometer. This project is in the conceptual phase, not yet approved by the ALICE Collaboration. The electromagnetic calorimeter would feature a very high granularity (cell size of ~ 1 mm^2), most likely in a SiW sandwich design. The same MAPS technology would be used for read-out as for the upgraded ITS and MFT, focusing on the good position resolution and two-particle separation. The electromagnetic calorimeter could be complemented by a hadronic calorimeter that would improve the jet reconstruction performance. The FoCal would allow detection of direct photons and neutral pions in a pseudo-rapidity and transverse momentum range that is not accessible for existing experiments, but where saturation effects are expected to appear [13].

4. Summary

The ALICE Upgrade program targets the LHC Long Shutdown 2 for detector installation to exploit the physics potential of the high luminosity LHC heavy-ion operations and to perform detailed and quantitative characterisation of the high density and temperature phase of strongly interacting matter, with a physics program extending beyond the LHC Long Shutdown 3.

[1] KAamodt *et al.* [ALICE Collaboration], Phys. Rev. Lett. **105** (2010) 252302 [arXiv:1011.3914 [nucl-ex]].
[2] B. Abelev *et al.* [ALICE Collaboration], Phys. Lett. B **719** (2013) 18 [arXiv:1205.5761 [nucl-ex]].
[3] K. Aamodt *et al.* [ALICE Collaboration], Phys. Lett. B **696** (2011) 30 [arXiv:1012.1004 [nucl-ex]].
[4] B. Abelev *et al.* [ALICE Collaboration], arXiv:1305.2707 [nucl-ex].
[5] B. Abelev *et al.* [ALICE Collaboration], JHEP **1209** (2012) 112 [arXiv:1203.2160 [nucl-ex]].
[6] B. Abelev *et al.* [ALICE Collaboration], Phys. Rev. Lett. **109** (2012) 072301 [arXiv:1202.1383 [hep-ex]].
[7] ALICE Collaboration, CERN-LHCC-2012-012 (2012).
[8] ALICE Collaboration, CERN-LHCC-2012-013 (2012).
[9] ALICE Collaboration, Upgrade of the ALICE Inner Tracking System, CERN-LHCC-2013-024, ALICE-TDR-017 (2013).
[10] K. Aamodt *et al.* [ALICE Collaboration], JINST **3** (2008) S08002.
[11] ALICE Collaboration, Addendum of the Letter Of Intent for the Upgrade of the ALICE Experiment : The Muon Forward Tracker, CERN-LHCC-2013-014; LHCC-I-022-ADD-1 (2013).
[12] A. Uras and f. t. A. M. W. Group, arXiv:1212.6517 [hep-ex].
[13] C. A. Salgado, J. Phys. G **38** (2011) 124036 [arXiv:1108.5438 [hep-ph]].

Measurements of jets in ALICE

Christine Nattrass for the ALICE collaboration
1408 Circle Drive, Knoxville, TN 37996, USA

E-mail: christine.nattrass@utk.edu

Abstract. Partonic energy loss in the medium formed in heavy ion collisions results in significant modifications of jet spectra. Quantitative understanding of these modifications can constrain models for partonic energy loss in heavy ion collisions. The ALICE detector is capable of unique measurements of jets due to its low momentum tracking and particle identification capabilities. The ALICE Electromagnetic Calorimeter (EMCAL) is a key element for the measurement of fully reconstructed jets in ALICE due to its measurement of neutral particles and its triggering capabilities. Measurements of fully reconstructed jet spectra and the nuclear modification factor, R_{AA}, in Pb-Pb collisions at $\sqrt{s_{NN}} = 2.76$ TeV are presented.

1. Introduction

Jets, the collimated sprays of particles created from a fragmenting parton, are an ideal probe of the hot, dense medium formed in high-energy heavy-ion collisions because hard parton scattering occurs early in the collision and therefore the scattered partons can interact with the medium. The production of jets is well understood in the absence of a medium, for example in pp collisions, and calculable in pQCD. Jets may be modified by the medium by parton energy loss through gluon brehmsstralung, collisional energy loss, or modification of fragmentation and hadronization. Studies of jets in heavy ion collisions at both the Relativistic Heavy Ion Collider (RHIC) and the Large Hadron Collider (LHC) have indicated that high momentum colored probes interact strongly with the Quark Gluon Plasma (QGP). The effects of parton interactions with the QGP have already been observed [1–6].

The ALICE detector provides unique capabilities due to its low momentum acceptance ($p_T >$ 150 MeV/c). This means that measurements of jets with the ALICE detector are less sensitive to corrections that are dependent on the shape of the fragmentation functions at low momenta in heavy ion collisions. The particle identification capabilities of the ALICE detector will also enable novel measurements of jets in heavy ion collisions.

2. The ALICE detector

The ALICE detector [7], shown in Figure 1, is designed for measurements of events with high track densities with precision detectors focused on studies around midrapidity ($|\eta| < 0.9$). The central detectors are in a 0.5 T magnetic field. The primary detectors used for the reconstruction of jets are the Inner Tracking System (ITS), the Time Projection Chamber (TPC), and the Electromagnetic Calorimeter (EMCAL) [8]. The TPC extends from a radius of approximately 85 cm to 250 cm from the beam pipe and when combined with the ITS provides tracking within $|\eta| < 0.9$ with momentum resolution $\Delta p_T/p_T$ ranging from approximately 1% below 10 GeV/c to 5% at 100 GeV/c for tracks completely contained in the TPC acceptance. The ITS surrounds

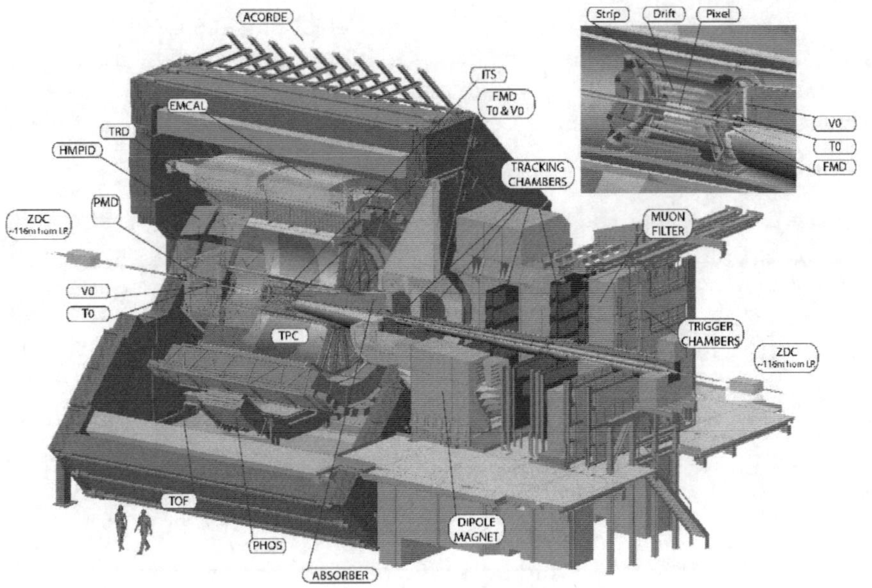

Figure 1. The ALICE detector.

the beam pipe and consists of a Silicon Strip Detector (SSD), Silicon Drift Detector (SDD), and Silicon Pixel Detector (SPD). The EMCAL is a lead scintillator sampling calorimeter covering $|\eta| < 0.7$ and 107° in azimuth optimized for studies of jets and capable of triggering on jets.

3. Measurements of jets

There are three steps in the jet reconstruction procedure: identification of jet candidates, background subtraction, and corrections for energy resolution and energy scale. Jet candidates are reconstructed using the anti-k_T algorithm in FastJet [9, 10]. ALICE reconstructs jets in two ways, using tracks only [6] and using information from the tracking detectors and from the EMCAL [11]. The former are called "charged jets" and the latter "full jets." For full jet reconstruction, tracks reconstructed using the TPC and ITS with $p_T > 150$ MeV/c and calorimeter clusters from the EMCAL with $E_{\text{cluster}} > 300$ MeV/c are used as input for the jet-finding algorithm. The boost-invariant p_T recombination scheme [9] is used, meaning that jet momentum $p_{T,jet}$ is the scalar sum of the constituent momenta. Charged particles will deposit some energy into the EMCAL. This energy needs to be removed to prevent double counting. Tracks are extrapolated to the EMCAL and then geometrically matched to clusters. The track momentum is subtracted from the matched clusters, and clusters that fall below 300 MeV after this correction are discarded. Jet candidates are required to be at least $R = \sqrt{\Delta\phi^2 + \Delta\eta^2}$ away from the EMCAL boundaries. For R=0.2 this corresponds to an acceptance of $\eta < 0.5$ in pseudorapidity and $1.60 < \phi < 2.94$ in azimuth.

The background subtracted jet p_T is given by

$$p_{T,jet} = p_{T,jet}^{\text{rec}} - \rho A_{\text{jet}} \tag{1}$$

where $p_{T,jet}^{rec}$ is the p_T reconstructed from the jet finder, ρ is the average energy density per unit area and A_{jet} is the area of the jet. For jet finding algorithms such as anti-k_T which reconstruct jets in cone symmetric in ϕ and η, $A_{jet} = \pi \text{R}^2$. Due to the restricted acceptance for full jets, ρ is calculated from the background from charged jets, ρ_{ch}. Charged jets are reconstructed using a

Journal of Physics: Conference Series **589** (2015) 012015 doi:10.1088/1742-6596/589/1/012015

random cone, the two jets with the highest p_T are excluded, and the median value of $p^{rec}_{T,jet}/A_{jet}$ is used to determine ρ_{ch} [5]. This is done event-by-event. To calculate ρ from ρ_{ch}, the ρ_{ch} is scaled up by a factor s, $\rho = \rho_{ch} * s$. This centrality-dependent factor is the ratio of the charged plus neutral energy to the charged energy in the event averaged over the event class determined from data.

The $p_{T,jet}$ from Eq.(1) still needs to be corrected for detector effects and smearing due to the fluctuations in the background. The response matrix which describes these effects and is used in unfolding is separated into a component due to background fluctuations and a component due to detector effects such as finite track reconstruction efficiency. The response matrix for detector effects is determined from simulated PYTHIA events run through GEANT. Particle level jets from the raw PYTHIA events are correlated with detector level jets from PYTHIA+GEANT and form the detector response matrix [12]. The fluctuations in the background are determined by the difference between the reconstructed energy and the average background energy given by

$$\delta p_T = p^{rec}_{T,\mathrm{RC}} - \rho\pi R^2 \tag{2}$$

where $R = \sqrt{\Delta\phi^2 + \Delta\eta^2}$ is the radius of the random cone. The δp_T distribution is calculated by placing cones randomly in the event, and summing up the energy within them, and subtracting the average energy that should be contained in a cone of that size. The width of this distribution indicates the size of the background fluctuations. This distribution is used to create the background response matrix. Unfolding is done using the Bayesian method in the RooUnfold package [13]. The uncertainty is dominanted by the uncertainty due to the tracking efficiency and the unfolding algorithm.

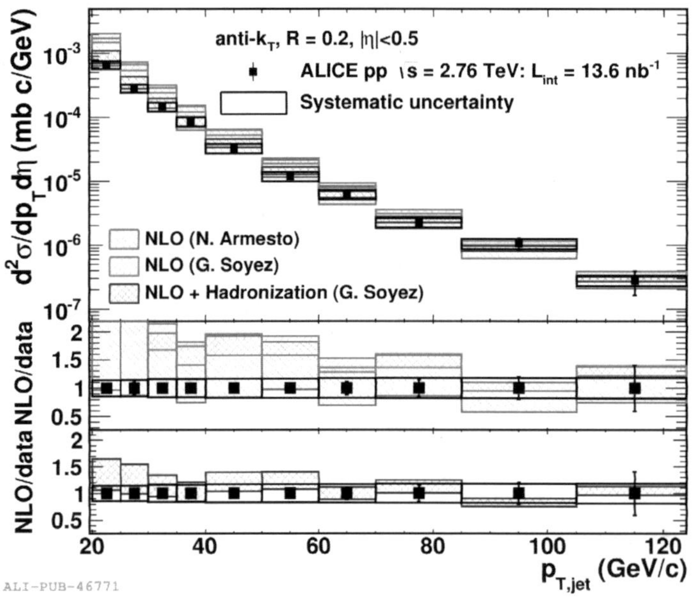

Figure 2. Inclusive p_T spectrum of fully reconstructed jets in pp collisions at $\sqrt{s} = 2.76$ TeV for the anti-k_T algorithm with R=0.2 compared to NLO calculations [11].

Measurements of full jets in pp collisions are consistent with pQCD calculations including hadronization for both R=0.2, show in figure 2, and R=0.4 [11], indicating that the measurement is well understood theoretically in pp collisions.

Journal of Physics: Conference Series **589** (2015) 012015 doi:10.1088/1742-6596/589/1/012015

To reduce the background in heavy ion collisions, the jet candidate sample is biased by requiring at least one track with $p_T > 5$ GeV/c. For calculations of the nuclear modification factor, R_{AA}, this bias is also used in pp collisions.

Full jet spectra in Pb-Pb collisions at $\sqrt{s_{NN}} = 2.76$ TeV are shown in figure 3 for R=0.2. The nuclear modification factor, R_{AA}, is shown in figure 4. Substantial suppression is observed, decreasing with p_T. These results are consistent with measurements from CMS [3], which were measured using a similar method, in the kinematic region where the results overlap. These results are also consistent with the charged particle spectra at the same momenta [14].

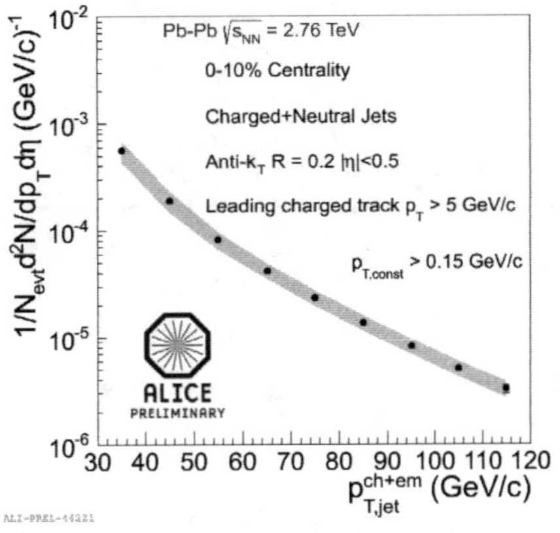

Figure 3. Inclusive spectrum of fully reconstructed jets in Pb-Pb collisions at $\sqrt{s_{NN}} = 2.76$ TeV for the anti-k_T algorithm with R=0.2.

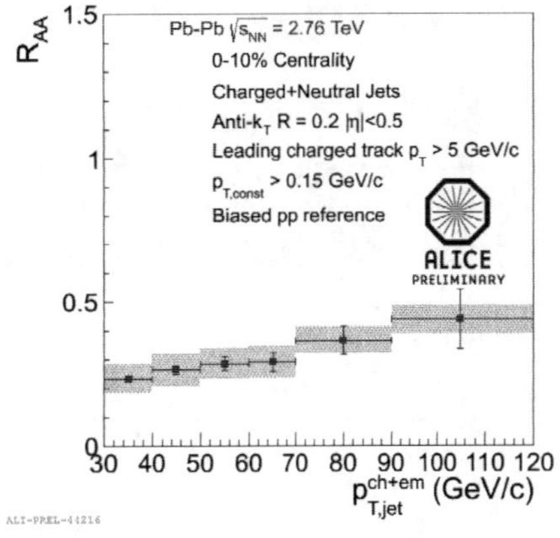

Figure 4. Inclusive spectrum of fully reconstructed jet R_{AA} in Pb-Pb collisions at $\sqrt{s} = 2.76$ TeV for the anti-k_T algorithm with R=0.2.

Future studies will incorporate the Di-jet Calorimeter (DCAL) [15] that is being installed 180° from the EMCAL and will cover $|\eta| < 0.7$ and 67° in azimuth. This will enable triggering on jets opposite to an energetic deposit such as that expected from a photon or high energy electron from a heavy flavor decay.

[1] Aad G *et al.* (ATLAS Collaboration) 2013 *Phys.Lett.* **B719** 220–241 (*Preprint* 1208.1967)
[2] Aad G *et al.* (ATLAS Collaboration) 2010 *Phys.Rev.Lett.* **105** 252303 (*Preprint* 1011.6182)
[3] Chatrchyan S *et al.* (CMS Collaboration) 2012 *Eur.Phys.J.* **C72** 1945 (*Preprint* 1202.2554)
[4] Chatrchyan S *et al.* (CMS Collaboration) 2013 (*Preprint* 1310.0878)
[5] Abelev B *et al.* (ALICE Collaboration) 2012 *JHEP* **1203** 053 (*Preprint* 1201.2423)
[6] Verweij M (ALICE Collaboration, Marta Verweij collaboration) 2012 (*Preprint* 1208.6169)
[7] Aamodt K *et al.* (ALICE Collaboration) 2008 *JINST* **3** S08002
[8] Cortese P *et al.* (ALICE Collaboration) 2008 ALICE Electromagnetic Calorimeter Technical Design Report Tech. Rep. CERN-LHCC-2008-014. ALICE-TDR-14 CERN Geneva
[9] Cacciari M, Salam G P and Soyez G 2012 *Eur.Phys.J.* **C72** 1896 (*Preprint* 1111.6097)
[10] Cacciari M and Salam G P 2006 *Phys.Lett.* **B641** 57–61 (*Preprint* hep-ph/0512210)
[11] Abelev B *et al.* (ALICE Collaboration) 2013 *Phys.Lett.* **B722** 262–272 (*Preprint* 1301.3475)
[12] Aiola S (ALICE) 2013 *J.Phys.Conf.Ser.* **446** 012005 (*Preprint* 1304.6668)
[13] Adye T 2011 6 p comments: 6 pages, 5 figures, presented at PHYSTAT 2011, CERN, Geneva, Switzerland, January 2011, to be published in a CERN Yellow Report
[14] Aamodt K *et al.* (ALICE Collaboration) 2011 *Phys.Lett.* **B696** 30–39 (*Preprint* 1012.1004)
[15] Allen J *et al.* 2010 ALICE DCal: An Addendum to the EMCal Technical Design Report Di-Jet and Hadron-Jet correlation measurements in ALICE Tech. Rep. CERN-LHCC-2010-011. ALICE-TDR-14-add-1 CERN Geneva

Journal of Physics: Conference Series **589** (2015) 012016 doi:10.1088/1742-6596/589/1/012016

Studying possible higher twist contributions in the inclusive charged hadron cross sections

Esko Pohjoisaho

Helsinki Institute of Physics, P.O. Box 64, FI-00014 University of Helsinki, Finland
Department of Physics, P.O. Box 35 (YFL), FI-40014 University of Jyväskylä, Finland

E-mail: esko.pohjoisaho@jyu.fi

Abstract.

In the standard pQCD picture particles are produced via the parton jet fragmentation process. However, there are also other production mechanisms like higher twist (HT) processes. A usual example of a HT process is a direct production of an outgoing hadron, where the hadron is produced in the hard subprocess without fragmentation.

We study the HT phenomena using a shape analysis (x_{T} scaling) of the inclusive invariant cross sections of charged hadrons, measured by the ALICE collaboration at center-of-mass energies $\sqrt{s} = 2.76$ TeV and 7 TeV. The data is compared to PYTHIA8 event generator and to a phenomenological model for HT. Using PYTHIA8, we explore a possible enhancement of HT phenomena for isolated particles, by comparing the shapes of the isolated distributions to inclusive distributions. The results from the standard PYTHIA8, without HT, is compared to a PYTHIA8 where we had included a HT process.

Finally, we found out that the effects observed in the x_{T} spectra originate from kinematic biases posed by the isolation cuts, rather than from an enrichment of the HT hadrons at the observed cross sections. A more detailed data analysis is ongoing.

1. Introduction

The cross section for inclusive pion production in pp collisions at $\sqrt{s} = 200$ GeV measured by RHIC [1] was described well by an NLO pQCD calculation for $p_{\mathrm{T}} > 2$ GeV/c. The agreement spanned seven orders of magnitude in the cross section. However, at the LHC energies, deviations of factor two between the NLO and data have been reported by the ALICE collaboration [2] and CMS [3] (see Fig 1). It is already known that selecting a different fragmentation function [4] for the NLO calculation can bring down the deviation with data to 25 % [5]. However, it is not yet fully understood to what extent the so called higher twist (HT) processes, additional to NLO, contribute to the cross section. The HT processes have attracted attention also in the field of heavy ion collisions, in the so-called "Baryon anomaly", found in measurements of high-p_{T} baryon production in heavy ion collisions [6]. So far the question of HT in the LHC is not settled [7].

1.1. x_{T} scaling

The kinematics of a $2 \rightarrow 2$ process is completely fixed by the two dimensional quantities $\sqrt{s}$ and p_{T}, and the two angles ϕ and θ. By using dimensional analysis and the scaled transverse

Journal of Physics: Conference Series **589** (2015) 012016 doi:10.1088/1742-6596/589/1/012016

momentum $x_T = \frac{2p_T}{\sqrt{s}}$, one can express the invariant cross section in a factorized form

$$E\,\frac{d^3\sigma}{dp^3} = \frac{1}{p_T^n} F\left(\frac{2p_T}{\sqrt{s}}, \theta\right) = \frac{1}{\sqrt{s}^n} G(x_T, \theta), \tag{1}$$

where F and G are dimensionless scaling functions. Any combination of p_T, $\sqrt{s}$ or the scaled transverse momentum x_T that preserve the dimensions, will result in the scaling form of Eq. (1). For any scale free theory and vector boson exchange process, the scaling exponent is $n = 4$. However, the experimentally measured n is not a constant $n = 4$, but depends on the studied process, collision energy and the range of x_T, reaching values up to $n \sim 8$ [8]. For example, the intrinsic partonic transverse momentum k_T, QCD radiation, running coupling α_s, scaling violations in parton distribution functions (PDF) and fragmentation functions (FF) and smearing from the jet fragmentation transverse momentum j_T lead into the experimentally observed $n > 4$ [8, 9]. For direct photons the exponent is predicted to be roughly one unit smaller due to the absence of fragmentation and one power less in α_s. This is also supported by the experimental data for direct photons and jets [8]. At the LHC energies the pQCD predicts $n \approx 5 \ldots 6$.

To explicitly study the behaviour of the scaling exponent, one can choose invariant cross sections from two data sets with different $\sqrt{s}$ at the same values of x_T. Using Eq. (1) for both cross sections allows to cancel out the dimensionless scaling functions to get

$$n(x_T, \sqrt{s}_1, \sqrt{s}_2) = \frac{\ln(\sigma^{\rm inv}(x_T, \sqrt{s}_2)/\sigma^{\rm inv}(x_T, \sqrt{s}_1))}{\ln(\sqrt{s}_1/\sqrt{s}_2)}. \tag{2}$$

An experimental demonstration of x_T scaling of charged hadron production in proton-proton collisions measured by the CMS collaboration at CERN and CDF collaboration in Fermilab [3], along with a global power law-fit, is presented in the upper panel of Fig. 1. Cross sections have been scaled by $\sqrt{s}^n$, where the scaling exponent is $n = 4.9$, and one can see curves with different center of mass energies collapse on top of each other in the high x_T range. The black points are for 7 TeV in CMS, red circles for 0.9 TeV in CMS, orange stars for 1.96 TeV in CDF, green crosses for 1.8 TeV CDF and yellow diamonds for 0.63 TeV CDF data. In the lower panel, ratios of data to NLO predictions are shown for different center-of-mass energies [3].

1.2. Higher twist

The term "twist" is a historical relic originating from the operator product expansion, which was a tool used to obtain perturbative predictions for deep-inelastic scattering [10]. Today the leading twist (LT) is understood as standard processes of the pQCD within the collinear factorization, where hadrons are produced via fragmentation processes. In contrast, HT processes are often understood as direct hadron production, where the hadron is produced directly in the hard subprocess without fragmentation [8]. Due to the lack of fragmentation in HT, a production of a direct hadron is argued to involve a large number of active fields, which, according to dimensional counting rules presented in [6], lead to steeper p_T spectra and larger scaling exponents n than for the LT hadrons. Therefore the scaling exponents are compared with those obtained from the NLO pQCD [5], as was done in [8].

The first calculations to estimate the size of the higher-twist contributions were carried out by Bagger and Gunion [11]. They found that the pion form factor in the $\gamma q \to \pi q$ process resulted in an extra factor of $1/\hat{s}$ where $\hat{s}$ is the partonic center-of-mass energy, causing an $n = 6$ scaling form for the cross section in Eq. (1). The HT contribution has been estimated to be significant especially in the kinematical range of high x_T and low p_T, i.e. at low $\sqrt{s}$ [10, 12].

Previously, there have been various attempts to experimentally measure the magnitude of the HT effects. Collaborations such as WA77, WA69 and the OMEGA were presenting their findings

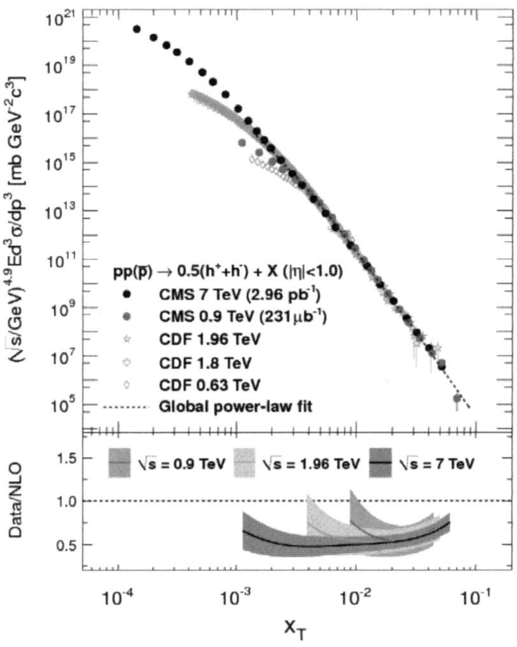

Figure 1. (Upper panel) Invariant cross sections as a function of x_{T} scaled by $(\sqrt{s})^{4.9}$ of inclusive charged particles to demonstrate x_{T} scaling with different $\sqrt{s}$. (Lower panel) Ratios between data and NLO QCD calculation is shown. Figure from [3].

of direct meson production in a Workshop on High p_{T} Physics and Higher Twists in 1988. At that time there were many promising results interpreted as evidence of HT, in processes such as $\bar{\nu}N \to \mu^+\pi^-X$, Drell-Yann, ρ production and in e^+e^- data [13].

In search for direct meson production in π^- Be interactions by the WA77 collaboration [14], the ρ^0 production was found to be consistent with LUCIFER Monte Carlo (MC) generator [15] with a small HT contribution, whereas the other mesons such as ϕ and K were found to be consistent with the purely leading twist MC. Based on MC studies of γq collisions carried out by the WA69 experiment [16], it was found that the shape of the inclusive p_{T} distribution was not changed significantly by the HT. Based on this remark it would be preferable to take advantage of the expected special kinematical properties of a HT process, by using cuts that enrich the fraction of HT particles in the data. Because the direct meson is created without fragmentation, one could argue that using an isolation cut would enrich the fraction of these mesons in the data, by suppressing the LT mesons originating from fragmentation processes. The effect of isolation was studied in the MC level by the OMEGA Photon Collaboration [13], where the meson was not allowed to have any track in its vicinity in a cone of a fixed size. The result was that this kind of isolation did not help significantly in the kinematical range where the HT processes were expected to be significant compared to LT. Still, by including a HT component on top of a QCD based model was in somewhat better agreement with the data than a fit without HT, leaving some room for HT contribution in the interpretation of the data.

Journal of Physics: Conference Series **589** (2015) 012016 doi:10.1088/1742-6596/589/1/012016

More recently, the x_{T} scaling exponents from world data were studied in [8]. Especially for hadrons at large x_{T}, the exponents were found to be systematically larger than predicted from the NLO pQCD calculations, whereas the photon and jet exponents were in agreement with the theory. The authors suggested that the presence of HT contributions could be a possible explanation for the large exponents as compared to the NLO calculation.

To create predictions for the scaling exponents at RHIC and LHC energies, a two component model was proposed [8] for the cross section

$$\sigma^{\mathrm{model}}(pp \to \pi\ \mathrm{X}) \propto \frac{A(x_{\mathrm{T}})}{p_{\mathrm{T}}^4} + \frac{B(x_{\mathrm{T}})}{p_{\mathrm{T}}^6}, \tag{3}$$

where functions $A(x_{\mathrm{T}})$ and $B(x_{\mathrm{T}})$ represent LT and HT contributions with their typical p_{T} dependence. Taking A and B as constants, the effective exponent in this model depends on the relative strength of the HT corrections to the LT cross section, reflected by the ratio B/A [8]. Using Eq. (3) and the NLO scaling exponent, they constructed an effective exponent

$$\begin{aligned} n(x_{\mathrm{T}}, p_{\mathrm{T}}, B/A) &\equiv -\frac{\partial \ln \sigma^{\mathrm{model}}}{\partial \ln p_{\mathrm{T}}} + n^{\mathrm{NLO}}(x_{\mathrm{T}}, p_{\mathrm{T}}) - 4 \\ &= \frac{2B/A}{p_{\mathrm{T}}^2 + B/A} + n^{\mathrm{NLO}}(x_{\mathrm{T}}, p_{\mathrm{T}}). \end{aligned} \tag{4}$$

In Eq. (4) the terms $n^{\mathrm{NLO}} - 4$ have been added to the logarithmic derivative so that the n converges into n^{NLO} when the HT contributions vanish (i.e. $B \to 0$). This corresponds to the definition of NLO pQCD as a LT in [8], although from Eq. (3) the effective exponent for LT contribution would be $n^{\mathrm{model}}_{B\to 0} = 4$ for fixed x_{T}, corresponding to a scale free theory. By fitting Eq. (4) to the world data, the authors of [8] obtained $B/A \sim 50\,\mathrm{GeV}^2$. By parametrization of Eq. (4) they created predictions for the scaling exponents at RHIC and LHC energies.

2. Inclusive x_{T} scaling exponents by ALICE

Differential cross sections of charged particles in inelastic pp collisions at $\sqrt{s} = 0.9, 2.76$ and 7 TeV have been measured by the ALICE collaboration at the LHC. The measurements were performed in the pseudorapidity range $|\eta| < 0.8$ for particles having $p_{\mathrm{T}} > 0.15$ GeV/c [2]. The data were collected based on tracking information from the Inner Tracking System (ITS) and the Time Projection Chamber (TPC), which are located in the central barrel of the experiment. The minimum-bias interaction trigger was derived using signals from the forward scintillators (VZERO), and the two innermost layers of the ITS, called the Silicon Pixel Detector (SPD) [2].

In this study, the cross sections as a function of x_{T} at 2.76 and 7 TeV were derived from the charged hadron p_{T} spectra measured by ALICE [2], and the scaling exponents were calculated using Eq. (2) for the x_{T} spectra. In Fig. 2 (a), the scaling exponents are plotted as a function of x_{T} for the NLO calculation [5] (black line), default PYTHIA8 simulation [17] (blue points), and for the exponents calculated from the inclusive charged particle data measured by the ALICE collaboration [2] (red points) at 7 TeV and 2.76 TeV.

In Fig. 2 (b) the NLO exponents have been subtracted from the ALICE data and PYTHIA8 exponents from Fig. 2 (a), to compare the results with prediction bands from [8], where the prediction was based on further parametrization of Eq. (3) for the difference to the NLO exponent, i.e. $\Delta^{\mathrm{fit}} = n_{\mathrm{measured}} - n_{\mathrm{NLO}}$. The red band is the prediction for the LHC energies and the blue band for RHIC energies [18]. The black points are preliminary data from the PHENIX collaboration [19], blue points from the default PYTHIA8 and red points the ALICE data.

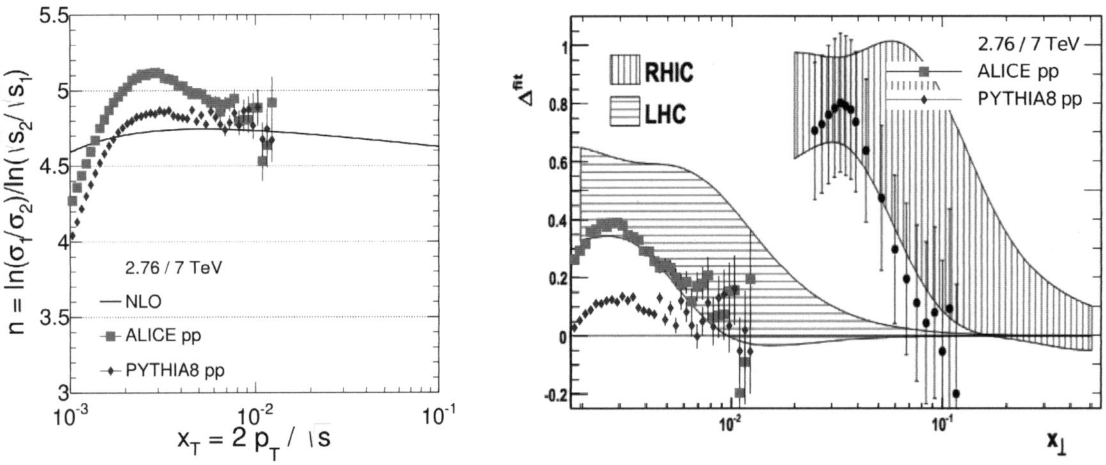

(a) Scaling exponents n (b) Difference to NLO, Δn

Figure 2. (a) Scaling exponents n extracted from NLO, ALICE [2] and PYTHIA8 pp data. (b) The difference Δn to the NLO exponents (i.e. Δ^{fit}) from ALICE, PHENIX preliminary data [19] and PYTHIA8, with prediction bands based on the two-component model [8].

The data points in Fig. 2 (b) seem to be in agreement with the prediction. Preliminary data points from PHENIX have been shown to be in agreement with the prediction band for RHIC energies [18], but until now the inclusive ALICE data points have not been shown on top of the prediction for LHC energies. The points from PYTHIA8 are closer to the NLO than the ALICE data, but still non-zero. In fact, it was seen that PYTHIA8 describes the measured shapes of the cross sections well but the absolute normalization disagrees with the ALICE data $20-25$ % (not shown here). If the p_{T} cross sections were manually rescaled closer to data, the resulting scaling exponents n would (naturally) agree with the measured data, i.e. the difference between the data and PYTHIA8 on Fig. 2 (b) originates from the fact that this PYTHIA tune misses the absolute normalization of the cross sections.

3. PYTHIA implementation of higher twist

There are no HT processes, like direct hadron production, implemented in the standard PYTHIA8 [17]. A process of this kind was added [20] into PYTHIA8 using the elementary cross section for a HT direct pion production process ($q_A g \to q_B \pi$) computed in [12]. The cross section is

$$\frac{\mathrm{d}\hat{\sigma}}{\mathrm{d}\hat{t}} = \frac{\pi\alpha_s^2}{\hat{s}^2}\frac{\alpha_s}{4\pi}\frac{s_0}{\hat{s}}F(\cos\theta), \tag{5}$$

where $s_0 = 16\pi^2 f_\pi^2$ is a higher twist scale, fixed by the pion weak decay constant f_π, and $F(\cos\theta)$ is the angular function for the c.m. scattering angle θ, $\cos\theta = 1 + 2\hat{t}/\hat{s}$, and has a form

$$F(z) = \frac{2}{27}(1-z)\left(1 + \frac{4}{(1+z)^2}\right). \tag{6}$$

The effects of using an isolation cut was studied with PYTHIA8. The isolation condition was checked for every particle by adding up the p_{T} of other final charged particles around it inside a cone of radius $R = \sqrt{\Delta\eta^2 + \Delta\phi^2} = 0.4$. To declare a particle as isolated, the p_{T} sum in the cone was required to be less than $10\,\%$ of the particle's p_{T}. It has been suggested that isolation cut would enrich the fraction of HT in the data. As a result, the scaling exponents n obtained from the isolated x_{T} spectra were anticipated to be larger than in the inclusive case [8].

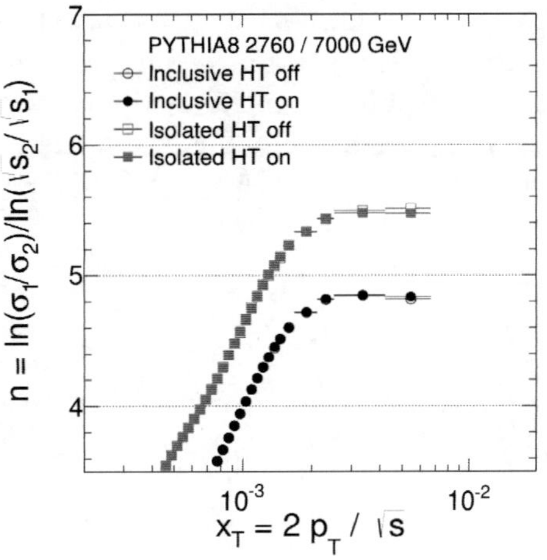

Figure 3. Scaling exponents from PYTHIA8 with Higher twist process on/off for inclusive and isolated hadrons.

In Fig. 3 one can observe that the effect of the HT process depicted by Eq. (5) to the exponent n in PYTHIA8 is quite negligible. In fact, there was on the average only one HT pion per 10000 charged hadrons in the simulation and applying the isolation cuts enriched the ratio only by a factor of 2. This suggests that the HT process described in [12] does not have a significant role in LHC energies.

One can see from Fig. 3 that the isolated spectrum has larger n than the inclusive spectrum, even without any HT processes, suggesting a kinematical bias caused by the isolation cut itself. This effect was further studied by a toy Monte Carlo model based on the "the parent-child relationship" (PCR), presented by Bjorken in [21]. Assuming that the fragmenting partons follow a power law distribution, $\mathrm{d}N/\mathrm{d}p_{\mathrm{Tq}} \sim p_{\mathrm{Tq}}^{-(p-1)}$, where the p_{Tq} is the parton p_{T}, the final hadron spectrum follows asymptotically the same shape

$$\frac{1}{p_{\mathrm{Th}}}\frac{\mathrm{d}N}{\mathrm{d}p_{\mathrm{Th}}} = \frac{1}{p_{\mathrm{Th}}^{p}}\int_{x_{\mathrm{T}}}^{1} z^{p-2}D(z)\mathrm{d}z, \tag{7}$$

where p_{Th} is the transverse momentum of the hadron and $D(z)$ is the fragmentation function. In the toy MC model we use the power law exponent $p = 6$ for the "2.76 TeV" and $p = 5$ for the "7 TeV" distributions, that roughly give the asymptotic behaviour of the measured data [2]. For the fragmentation function, we took for simplicity $D(z) \sim \exp(-8.2z)$, as was done in [22].

To get a simple estimate of the isolation effects corresponding 10 % cone activity, we first force $z > 0.9$ in the integral in Eq. (7), which yields a constant isolation fraction

$$\frac{\text{isolated}}{\text{inclusive}} = \frac{\int_{0.9}^{1} z^{p-2}D(z)\mathrm{d}z}{\int_{x_{\mathrm{T}}\approx 0}^{1} z^{p-2}D(z)\mathrm{d}z}. \tag{8}$$

By inserting fractions obtained from Eq. (8) into Eq. (2), we found out that the scaling exponent is increased by 0.77 as compared to the non-isolated case. In addition, a cascade process was used. By repeatedly sampling a fragmentation variable z from the exponential "fragmentation function", one could create a collection of hadrons from the initial parton. The isolation condition for the leading hadron could then be checked by comparing the hadron p_{T} to the p_{T} sum of all the other hadrons in the toy MC event. It was observed that mostly the high z hadrons survive the isolation cut. Isolation probabilites obtained with this method are shown in Fig. 4 (a), and the increase of the scaling exponents, $\Delta n = n_{\mathrm{isolated}} - n$, is shown in Fig. 4 (b). Using the isolation fractions from Fig. 4 (a) with Eq. (2), the scaling exponent increased by 0.679 ± 0.008, marked by the red line in Fig. 4 (b). This result is of similar magnitude as with the integral method. More importantly, the increase of the exponent in the toy MC model is very close to the increase we observed in the standard PYTHIA8, see Fig. 3.

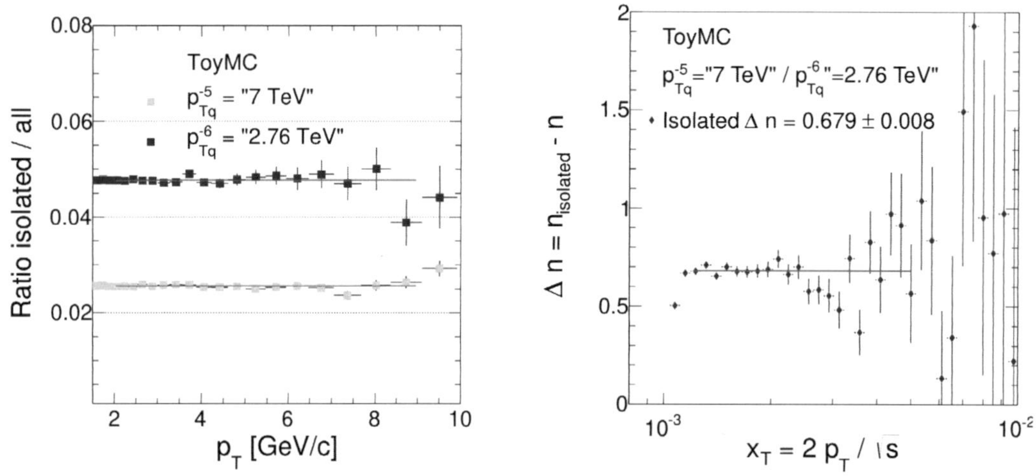

(a) Isolation fractions

(b) Increase of the scaling exponent n

Figure 4. (a) Isolation fractions obtained from the toy MC model for a parton cascade. (b) The difference Δn caused by the isolation cut in the toy MC.

4. Conclusion

The x_{T} scaling exponents n for inclusive charged hadrons were extracted from ALICE pp data, PYTHIA8 and NLO at $\sqrt{s} = 7$ TeV and 2.76 TeV. The scaling exponents obtained from the ALICE data are in agreement with the two-component model predictions. It was shown that PYTHIA8 describes the shape of the p_{T} spectra quite well, but the normalization is off by $\sim 20\,\%$. If this normalization were correct, the scaling exponents would be similar to data. As it was expected that the HT contribution would steepen the spectra and thus increase n, the similar shape of the data with PYTHIA8 may suggest that the HT contribution is not very large.

An MC level study of HT contribution was made by implementing a direct $\pi^{\pm}$ production process [12] into PYTHIA8 [20]. It turned out that the fraction of HT pions of all charged hadrons was roughly 1/10000 in the simulation, and applying the isolation cuts did not significantly enrich the ratio. Hence, the direct pion production as described by [12] gives a negligible contribution to the cross section and leads into barely visible effects in both inclusive and isolated scaling exponents.

The isolation cut was seen to increase n due to a kinematical bias in two ways. Firstly, by biasing the fragmentation into large values of z and thus making the spectra steeper. Secondly, the p_{T} cross section is harder at 7 TeV than at 2.76 TeV, which lead into higher probability for a hadron to be isolated at 2.76 TeV collision energy. The difference in the isolation probabilities directly results in an increase of the scaling exponent. It seems that the observed effect could be largely explained by these kinematical biases. A more careful study would be needed to verify to

Journal of Physics: Conference Series **589** (2015) 012016 doi:10.1088/1742-6596/589/1/012016

what extent there is room for HT contributions in the data. Also, the isolation study is so far done only with PYTHIA8, while the analysis for the ALICE data is ongoing.

Acknowledgments

We are most grateful to Torbjörn Sjöstrand for his kind help in implementing the direct hadron production process into PYTHIA8. Special thanks to Hannu Paukkunen from the University of Jyväskylä for providing us the NLO pQCD cross section spectra used in this study. Finally, thanks to the Helsinki Institute of Physics for funding the research and making it possible to participate in the 9th International Workshop on High-pT Physics.

References

[1] Adare A *et al.* (PHENIX Collaboration) 2007 *Phys.Rev.* **D76** 051106 (*Preprint* 0704.3599)
[2] Abelev B B *et al.* (ALICE Collaboration) 2013 *Eur.Phys.J.* **C73** 2662 (*Preprint* 1307.1093)
[3] Chatrchyan S *et al.* (CMS Collaboration) 2011 *JHEP* **1108** 086 (*Preprint* 1104.3547)
[4] Kretzer S 2000 *Phys.Rev.* **D62** 054001 (*Preprint* hep-ph/0003177)
[5] Paukkunen H 2013 Private communication
[6] Brodsky S J and Sickles A 2008 *Physics Letters B* **668** 111 – 115 ISSN 0370-2693
[7] Rak J and Tannenbaum M J 2013 *High p_T physics in the heavy ion era.* Cambridge monographs on particle physics, nuclear physics and cosmology (Cambridge Univ. Press)
[8] Arleo F, Brodsky S J, Hwang D S and Sickles A M 2010 *Phys.Rev.Lett.* **105** 062002 (*Preprint* 0911.4604)
[9] Tannenbaum M J 2005 *J.Phys.Conf.Ser.* **27** 1–10 (*Preprint* nucl-ex/0507020)
[10] Owens J 1987 *Rev. Mod. Phys.* **59** 465–503
[11] Bagger J and Gunion J 1982 *Phys.Rev.* **D25** 2287
[12] Berger E L, Gottschalk T and Sivers D 1981 *Phys. Rev. D* **23**(1) 99–113
[13] Apsimon R *et al.* (OMEGA Photon Collaboration) 1991 *Z.Phys.* **C50** 179–184
[14] Bellogianni A *et al.* (WA77 Collaboration) 1996 *Z.Phys.* **C69** 235–242
[15] Ingelman G and Weigend A 1987 *Comput.Phys.Commun.* **46** 241
[16] Hoeger C (WA69 Collaboration) 1989 *Nucl.Phys.Proc.Suppl.* **7B** 255–262
[17] Sjostrand T, Mrenna S and Skands P Z 2008 *Comput.Phys.Commun.* **178** 852–867 (*Preprint* 0710.3820)
[18] Arleo F, Brodsky S J, Hwang D S and Sickles A M 2010 *Nuclear Physics B - Proceedings Supplements* **207–208** 81 – 84
[19] Bazilevsky A On behalf of the PHENIX collaboration at APS10 meeting http://meetings.aps.org/Meeting/APR1, slides http://www.phenix.bnl.gov/WWW/talk/newtalk.php?cid=APS10
[20] Sjöstrand T 2013 Private communication
[21] Bjorken J 1973 *Phys.Rev.* **D8** 4098
[22] Adler S *et al.* (PHENIX Collaboration) 2006 *Phys.Rev.* **D74** 072002 (*Preprint* hep-ex/0605039)

Journal of Physics: Conference Series **589** (2015) 012017 doi:10.1088/1742-6596/589/1/012017

A Brief Review of CMS Jet Measurements

Sevil Salur for the CMS Collaboration

Department of Physics and Astronomy Rutgers, the State University of New Jersey 136 Frelinghuysen Road Piscataway, NJ 08854-8019 USA

E-mail: salur@physics.rutgers.edu

Abstract. We present an overview of the current status of the recent jet measurements in PbPb and pp collisions at a nucleon-nucleon center-of-mass energy of 2.76 TeV that are studied with the CMS detector at the LHC. Jets are reconstructed using the anti-k_T sequential reconstruction algorithm on particle flow objects. To distinguish between different possible energy loss mechanisms and to provide an important experimental check on the validity of the assumed jet fragmentation, inclusive jets, flavor tagged jets and jet shapes are studied.

1. Introduction

During the last three decades, relativistic heavy ion collisions have been studied to map out the phase diagram of quantum chromodynamics (QCD) matter at increasing center-of-mass energies of < 5 GeV at the Alternating Gradient Synchrotron in Brookhaven, < 20 GeV at the the Super Proton Synchrotron at CERN and ≤ 200 GeV at the Relativistic Heavy Ion Collider (RHIC) in Brookhaven. The Large Hadron Collider (LHC) at CERN was commissioned at the end of 2009 and started delivering pp collisions at a center-of-mass energy of 7 TeV. In 2010 and 2011, LHC also delivered the heavy ion collisions (PbPb) at a center-of-mass energy of 2.76 TeV per nucleon pair along with reference data of pp collisions at the same center of mass energy. This was followed by a pPb run at 5.02 TeV together with a higher statistics pp run at 2.76 TeV in early 2013. A center-of-mass energy of 5 TeV per nucleon pair for heavy ion collisions at LHC is expected to be delivered starting in 2015.

Well identified decay products of partonic interactions at large momentum transfers, also called hard probes, are used to study the structure and dynamics of the QGP [1, 2, 3]. These probes are well calibrated, as their expected yields are calculable using the perturbative QCD (pQCD) theoretical framework and as their propagation through the medium is affected by strong interactions. Also, due to their short production time ($\tau \sim 1/p_T \leq 0.1$ fm/c), they can be used to identify the properties of the medium starting from the early stages of the collision. Their interactions within the hot QCD medium can be measured experimentally via the modification of the internal structure of jets by medium induced soft gluon radiation [4]. At RHIC, indirect measurements of energy loss in the medium ("jet quenching") have been made via observables of leading fragments of jets and their correlations [5, 6, 7]. These techniques were developed in order to suppress the large underlying event backgrounds in heavy ion collisions. While this approach has provided significant insights, including the color opaqueness of the QGP, these findings are only qualitative and are limited by the intrinsic biases of these measurements. For example, the leading hadron measurements are preferentially from the population of jets that has the least interaction with the medium. These measurements are also not sufficient to discriminate

Journal of Physics: Conference Series **589** (2015) 012017 doi:10.1088/1742-6596/589/1/012017

quantitatively between partonic energy loss formalisms and to extract key parameters such as the transport coefficient of the hot QCD medium to measure the stopping-power of the QGP precisely [8]. Measurements of jet structure and its modification in terms of energy flow therefore provides a much closer connection to the underlying theory.

Jets, the most common hard probe of QCD, are collimated sprays of hadronic decay products of hard-scattered partons. Measuring jets above the complex heavy ion background is a challenging task. With the order of magnitude increase in the centre-of-mass energy compared to RHIC, jets with energies well above the background from the underlying events are observed at the LHC. Numerous numbers of jet pairs appeared to be not balanced indicating strong jet quenching in PbPb collisions at $\sqrt{s_{NN}} = 2.76$ TeV. An example of a typical unbalanced di-jet event reconstructed in central PbPb collisions collected by the CMS experiment is shown in Figure 1 [9].

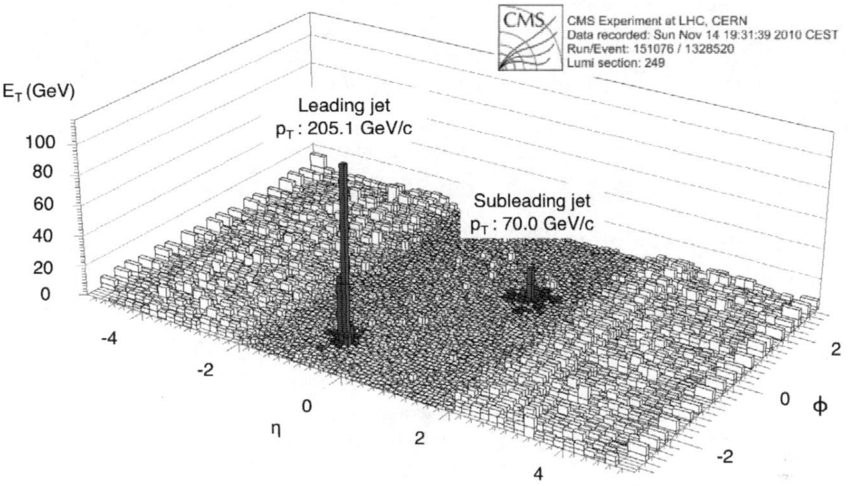

Figure 1. An example of a jet pair in a central PbPb collision event at $\sqrt{s_{NN}} = 2.76$ TeV. Plotted is the summed transverse energy in towers of electromagnetic and hadron calorimeters vs. η and ϕ, with the identified jets highlighted in red and labeled with the corrected jet transverse momentum [9].

2. Jet Measurements

For jet analyses presented here, a total integrated luminosity of 150 μb^{-1} PbPb collisions and a total integrated luminosity of 5.3 pb^{-1} of pp collisions at a nucleon-nucleon center-of-mass energy of 2.76 TeV, collected with the Compact Muon Solenoid (CMS) detector are used. The CMS experiment at the LHC is a general multi-purpose detector designed to explore physics at the large TeV energy scales [10]. The two most important detector requirements for a successful reconstruction of jets in heavy ion collisions are a good energy measurement and an efficient jet trigger. With its high quality electromagnetic and hadronic calorimeters covering a wide pseudorapidity and a full azimuthal range, by design CMS is extremely well suited to measure hard scattering processes [11]. Sub-detectors such as the high precision silicon tracker, which has very good momentum resolution, complement the calorimeters for jet studies.

Jet reconstruction in heavy-ion collisions in CMS is performed with anti-k_T sequential reconstruction algorithm that is encoded in the FastJet framework [12]. A small value of 0.3 for resolution parameter R is selected to reduce the deterioration of the jet energy resolution in PbPb collisions due to fluctuations of the background from soft interactions. An algorithm that is a variant of an iterative "noise/pedestal subtraction" technique is used to estimate the heavy ion background event-by-event [13]. The input particles of the jet reconstruction are Particle Flow (PF) objects that are reconstructed by combining information from various sub-detectors, most importantly by combining tracks with clusters in electromagnetic and hadronic calorimeters [14]. The reconstructed jet energies are corrected by using a factorized multi-step approach used for all jet analyses in CMS [15]. Jet energy corrections are derived from PYTHIA [16] simulations without PbPb underlying events.

2.1. Jet-Jet Pairs
By utilizing back to back jet pairs, it is possible to study modification of the jet properties such as back-to-back alignment of hard scattered partons, their energy balance, and even their fragmentation. An example of such study can be seen in Figure 8 of Ref. [9] that shows distributions of $\Delta\phi$ between leading and sub-leading jets and Figure 10 of Ref. [9] that shows the evolution of the diet momentum balance. In these measurements, angular correlation of jet pairs appear to be unmodified by the medium although a significant jet quenching at LHC independent of jet transverse momentum is observed [9, 17]. With the recent data collected earlier this year, jet pair production is also studied in pPb collisions at $\sqrt{s_{NN}} = 5.02$ TeV. In these collisions the pseudorapidity of the dijet system appears to change monotonically with increasing forward calorimeter activity consistent with expectations based on nuclear parton distribution effects [18, 19].

Studies of jet-hadron correlations involving vector summation of charged hadron momenta, find that the energy balance in events with large dijet asymmetry is recovered on average by an excess of low-momentum particles in the hemisphere of the away-side jet, at large angles relative to the jet axes [9]. Complementary information about the overall momentum balance in the dijet events can be obtained using the projection of missing p_T of reconstructed charged tracks onto the leading jet axis. These measurements reveal that the missing energy is also recovered in the form of soft particles at large angle with respect to the jet axes.

2.2. Boson Tagged Jet Pairs
While back-to back jet pair measurements confirm the jet quenching at LHC, these measurements introduce a selection bias towards scatterings occurring at, and oriented tangential to, the surface of the medium. A more complete measurement could be the so-called "Golden Probe" for jet tomography of the QGP, such as the Z^0 + jet and the photon+jet channel [20, 21, 22]. These channels have served as an essential calibrator of jet energy in TeV p+p collisions. In heavy ion collisions, they can again be used to calibrate in-medium parton energy loss, as they carry no color charge and escape the medium unattenuated.

As an example to show CMS detector capabilities, the absolute jet energy response measurement from the Z^0+jet events from the integrated luminosity of 36 pb^{-1} $\sqrt{s} = 7$ TeV p+p collisions is presented in the left panel of Figure 2. A good agreement between data and simulation is observed showing a good overall understanding of the detector. The invariant mass spectra of Z^0 bosons from $\sqrt{s_{NN}} = 2.76$ TeV PbPb collisions collected during the first heavy ion run is shown in the right panel of Figure 2. The shape of the spectrum is compared with the results from pp collisions, and the distribution is found to be consistent. Within uncertainties, no modification is observed with respect to theoretical next-to-leading order pQCD proton-proton cross sections scaled by the number of elementary nucleon-nucleon collisions. This measurement confirms the validity of the Glauber scaling for perturbative cross sections in nucleus-nucleus collisions at the LHC and establishes the feasibility of carrying out detailed Z^0+jet physics studies in heavy-ion collisions with the CMS detector.

However due to the limited available statistics of the collected data, the coincidence of Z^0+jet measurement is only feasible after 2015 with expected $\sim$2000 Z^0+jet coincidence counts that

Journal of Physics: Conference Series **589** (2015) 012017 doi:10.1088/1742-6596/589/1/012017

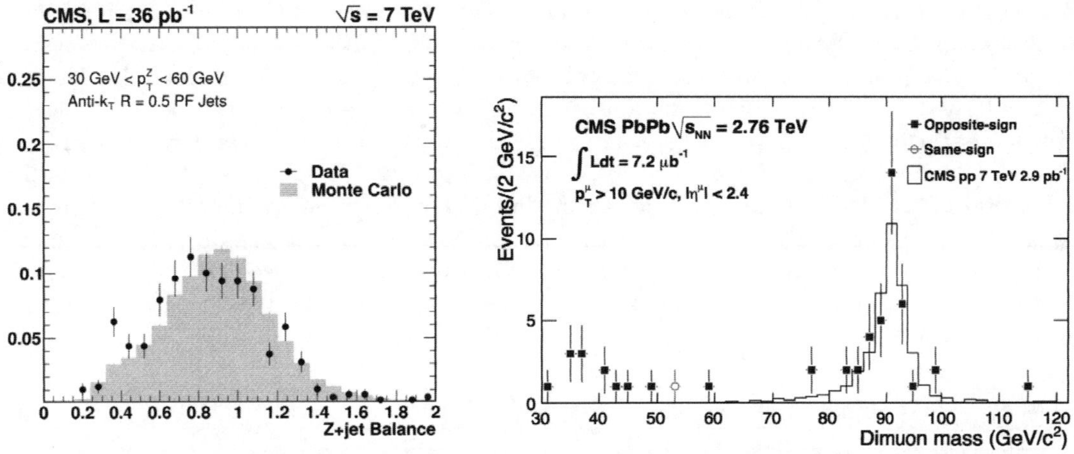

Figure 2. Left: jet energy response from $Z(\mu^+\mu^-)$+jets p_T-balancing for p+p collisions at $\sqrt{s} = 7$ TeV [23]. Right: Dimuon invariant mass spectra from $\sqrt{s_{NN}} = 2.76$ TeV PbPb collisions collected during the first heavy ion run [24]. Full squares are opposite-sign dimuons, while the empty circle shows a like-sign dimuon candidate. The histogram shows the corresponding distribution measured in pp collisions at 7 TeV scaled to the PbPb candidates.

decay into dimuons within the CMS acceptance from the expected one month of $\sqrt{s_{NN}} = 5$ TeV Pb+Pb run [11]. It is possible to perform photon+jet coincidence measurements with the available statistics [25]. In these measurements, the jet appears to be quenched but still back-to-back to its γ partner. This implies that jet energy is not lost in single hard gluon-radiation. Similar to jet pair coincidence measurements, energy loss observed also depends on centrality of the PbPb collisions.

2.3. Inclusive Jet Measurements

To gain sensitivity to the properties of the QGP and to quantify the underlying mechanism of the quenching effects, the ratios of inclusive jet cross-sections in PbPb collisions over the ones in pp are studied via the nuclear modification factors. If the full jet energy is recovered independent of the fragmentation details, even in the presence of strong jet quenching, this ratio should be equivalent to one. If the ratio is less than one, it would suggest broadening of the jet structure due to jet-medium interactions. The preliminary nuclear modification results show a suppression factor of 0.5 for high p_T jets in central PbPb collisions in comparison to pp collisions [26]. Also, no strong dependence on the jet radius is observed, unlike the predictions from theoretical NLO calculations that demonstrate the sensitivity of nuclear modification factors of different jet definitions to the properties of the medium-induced gluon radiation [27]. Other jet-medium interactions, initial state or cold nuclear matter effects might be affecting the interpretation of these measurements [28, 29, 30, 31]. The measurement of nuclear modification factors of jets in pPb collisions at $\sqrt{s_{NN}} = 5$ TeV will be performed to disentangle some of these effects on jet production.

Jet shape measurements describe how the jet transverse momentum is distributed as a function of the radial distance from the jet axis and are complementary to nuclear modification factor measurements. They are also robust against the elimination of low momentum particles, which can also be used to suppress and study heavy ion event fluctuations. The jet shape measurements are expected to be affected by the QGP induced gluon emission through substantial broadening of jets in medium relative to vacuum. They contain discriminative

Journal of Physics: Conference Series **589** (2015) 012017 doi:10.1088/1742-6596/589/1/012017

information on the energy loss mechanism to help characterize the parton-medium interactions such as transport coefficients [27, 33]. An example of differential jet shape measurements in PbPb and pp collisions as a function of the distance from the jet axis for inclusive jets with jet $p_T > 100$ GeV/c in various PbPb centrality intervals is presented in the top panel of Figure 3. To see the effect of the medium on jet shapes more clearly, the ratios of differential jet shapes in PbPb and pp collisions are also presented in the Figure 3. Deviations from unity indicate modification of jet structure in the nuclear medium. For example in the most central PbPb collisions an excess of transverse momentum fraction emitted at large radius $r > 0.2$ emerges, indicating a moderate broadening of the jets in the medium. This is also consistent with the jet fragmentation measurements [35] that indicate a transport of energy from intermediate to low p_T jet constituents.

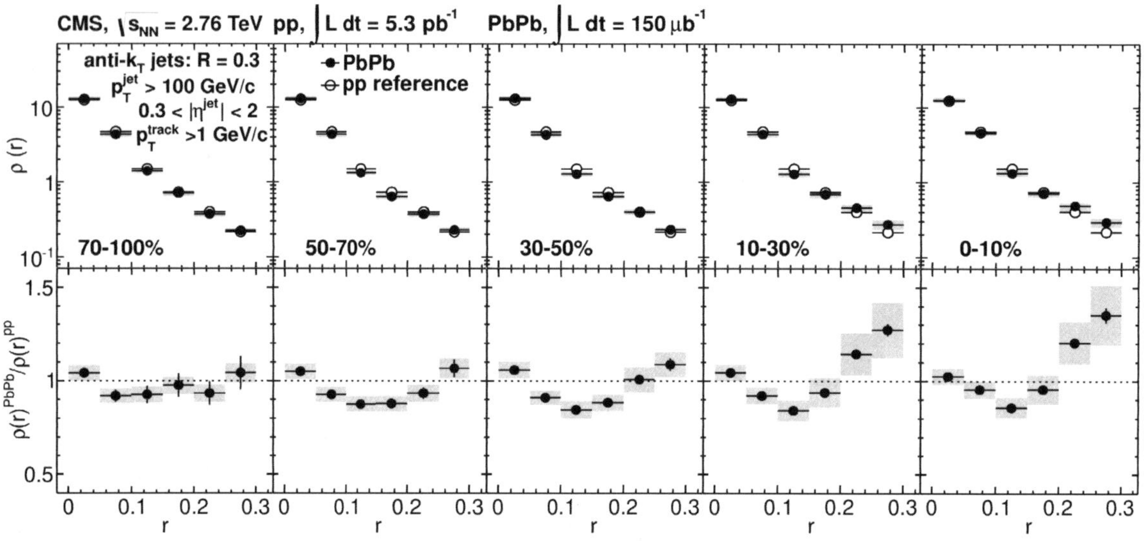

Figure 3. Top panel: Differential jet shapes in PbPb and pp collisions as a function of distance from the jet axis for various PbPb centrality intervals. Bottom panel: The ratio of jet differential jet shapes in PbPb and pp collisions [34].

2.4. Flavor Tagged Jets

The quenching of jets in heavy-ion collisions is expected to depend on the flavor of the initiating parton. For example, under the assumption that radiative energy loss is the dominant mechanism, gluon jets should be quenched more strongly than light quark jets, due to their larger color factor for gluon emission. On the other hand, jets initiated by heavy quarks, particularly bottom, are expected to radiate less than light ones. This is due to the so-called dead cone effect whereby radiation is suppressed in the direction of propagation as a result of coherence effects [36]. Recent CMS data demonstrates that non-prompt J/ψ, i.e., those coming from decay of B mesons, are indeed suppressed in PbPb collisions with respect to the pp expectation [37]. This indirect measurement, which samples typical J/ψ p_T values on the order of 10 GeV/c, provides a strong motivation to perform a more direct measurement using fully reconstructed jets. Such a measurement would enable a direct comparison of b quark energy loss to that of inclusive jets at much larger values of jet p_T, where the flavor dependence of parton energy loss can be probed in detail.

Journal of Physics: Conference Series **589** (2015) 012017 doi:10.1088/1742-6596/589/1/012017

Jets formed from heavy flavor quark fragmentation can be tagged by the presence of displaced vertices, either by direct reconstruction of these vertices or by the impact parameter (i.e., the distance of closest approach to the primary vertex) of tracks originating from these vertices [38]. Information from these tracks and vertices are typically combined into a quantity which optimizes their discrimination between heavy and light flavor jets. A discriminator based on the flight distance significance of reconstructed secondary vertices as the primary b-jet tagger is implemented in heavy ion collisions [39]. The b-jet coincidence measurement from the jet triggered $\sqrt{s_{NN}} = 2.76$ TeV Pb+Pb collisions were performed for the first time in heavy ion collisions [40, 41, 42].

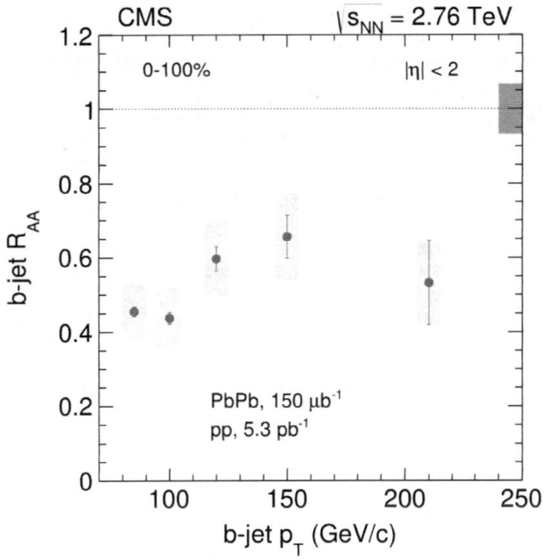

Figure 4. The centrality integrated b-jet nuclear modification factor as a function of jet p_T [40].

See Figure 4 for the centrality integrated b-jet nuclear modification factor, which is the ratio of PbPb yield scaled with the number of nucleon-nucleon collisions equivalent of pp and the measured pp cross section. The suppression of b-jets appears to be qualitatively consistent with that of inclusive jets [26]. The absence of a strong dependence of the jet suppression on the mass of the fragmenting parton is surprising. However a more detailed understanding of the parton mass and flavor dependence of energy loss would require precision studies such as back-to-back jet pairs to eliminate the contribution of b jets from gluon spitting and fragmentation functions to constrain the dynamics of heavy quark energy loss.

3. Conclusions

The quenching of jets in PbPb collisions is studied by complimentary jet measurements reconstructed in data collected by the CMS experiment. The anti-k_T sequential reconstruction algorithm is used to reconstruct jets based on combined tracker and calorimeter information. A strong increase in the fraction of jet pairs with very unbalanced transverse momentum is observed in central PbPb collisions compared to peripheral collisions while angular correlations observed to be conserved. The missing energy is also recovered in the form of soft particles at large angles with respect to the jet axis. A suppression factor of 0.5 for high p_T jets is observed in central PbPb collisions in comparison to pp collisions. A centrality dependent modification of the jet shapes and fragmentation functions appears in the central PbPb collisions. Through studies of bottom quark tagged jets, it appears that jet quenching does not have a strong dependence on parton mass and flavor in the jet. These results are all consistent and important for characterizing the jet quenching and shower evolution in the presence of a hot and dense nuclear medium.

The age of quantitative jet tomography has just begun, but we have yet to characterize the medium parton interactions in detail. High-precision studies of jets are expected to be possible in upcoming LHC running periods. We will measure complementary and robust jet observables in order to distinguish between competing energy loss mechanisms to determine key features of QCD and to extract quantitative properties such as QGP density and the corresponding stopping-power.

4. Acknowledgments

This material is based upon work supported by the National Science Foundation under Grant No. 1067907.

References

[1] M. Gyulassy, I. Vitev, X. N. Wang, B. W. Zhang, "Jet quenching and radiative energy loss in dense nuclear matter", Quark Gluon Plasma P123-191 (2003), [nucl-th/0302077].

[2] M. Gyulassy and M. Plumer, "Jet quenching as a probe of dense matter", Nucl. Phys. A 527 (1991) 641.

[3] X. N. Wang and M. Gyulassy, "Gluon shadowing and jet quenching in A + A collisions at $\sqrt{s_{NN}}$ = 200-GeV", Phys. Rev. Lett. 68 (1992) 1480.

[4] U. A. Wiedemann, "Jet Quenching in Heavy Ion Collisions", CERN-PH-TH-2009-026, (2009), [arXiv:0908.2306].

[5] J. Adams et al., [STAR Collaboration], "Evidence from d + Au measurements for final state suppression of high p_T hadrons in Au+Au collisions at RHIC", Phys. Rev. Lett. 91 (2003) 072304, [arXiv:nucl-ex/0306024].

[6] J. Adams et al., [STAR Collaboration], "Direct observation of dijets in central Au+Au collisions at $\sqrt{s_{NN}}$ = 200-GeV", Phys. Rev. Lett. 97 (2006) 162301, [arXiv: nucl-ex/0604018].

[7] S. S. Adler et al., [PHENIX Collaboration], "Common suppression pattern of η and π^0 mesons at high transverse momentum in Au+Au collisions at $\sqrt{s_{NN}}$ = 200-GeV", Phys. Rev. Lett. 96 (2006) 202301, [arXiv: nucl-ex/0601037].

[8] S. A. Bass, C. Gale, A. Majumder, C. Nonaka, G. Y. Qin, T. Renk and J. Ruppert, " Systematic Comparison of Jet Energy-Loss Schemes in a realistic hydrodynamic medium", Phys. Rev. C 79 (2009) 024901, [arXiv:0808.0908].

[9] CMS Collaboration, "Observation and studies of jet quenching in PbPb collisions at nucleon-nucleon center-of-mass energy = 2.76 TeV", Phys. Rev. C84, (2011) 024906, [arXiv:1102.1957].

[10] CMS Collaboration, "The CMS experiment at the CERN LHC", JINST S08004 (2008).

[11] CMS Collaboration, "CMS Physics Technical Design Report: Addendum on High Density QCD with Heavy Ions", J. Phys. G: Nucl. Part. Phys. 34 (2007) 2307.

[12] M. Cacciari, G. P. Salam and G. Soyez, "The anti-kt jet clustering algorithm", JHEP 0804 (2008) 063, [arXiv:0802.1189].

[13] O. Kodolova, I. Vardanian, A. Nikitenko et al., "The performance of the jet identification and reconstruction in heavy ions collisions with CMS detector", Eur. Phys. J. C 50 (2007) 117.

[14] CMS Collaboration, "Commissioning of the Particle-Flow Reconstruction in Minimum-Bias and Jet Events from pp Collisions at 7 TeV", CMS Physics Analysis Summary CMS-PAS-PFT-10-002, (2010).

[15] CMS Collaboration, "Determination of Jet Energy Calibration and Transverse Momentum Resolution in CMS", JINST 06 (2011) P11002, [arXiv:1107.4277].

[16] T. Sjostrand, S. Mrenna, and P. Skands, "PYTHIA6.4 Physics and Manual", JHEP 05 (2006) 026, [arXiv:hep-ph/0603175].

[17] CMS Collaboration, "Jet momentum dependence of jet quenching in PbPb collisions at $\sqrt{s_{NN}}$=2.76 TeV ", Phys. Lett. B 712 (2012) 176.

[18] CMS Collaboration, "Studies of dijet pseudorapidity distributions and transverse momentum balance in pPb collisions at sqrt(s[NN])=5.02 TeV", [arXiv:1401.4433] (2014).

[19] D. C. Gulhan for the CMS Collaboration, "Dijet Measurements in pPb collisions", CMS-CR-2013-463, CERN-CMS-CR-2013-463.

[20] X.-N. Wang, Z. Huang, and I. Sarcevic, "Jet Quenching in the Direction Opposite to a Tagged Photon in High-Energy Heavy-Ion Collisions", Phys. Rev. Lett. 77 (1996) 231, [arXiv:9701227].

Journal of Physics: Conference Series **589** (2015) 012017 doi:10.1088/1742-6596/589/1/012017

[21] X.-N. Wang and Z. Huang, "Medium-induced parton energy loss in γ+jet events of high-energy heavy-ion collisions", Phys. Rev. C 55 (1997) 3047, [arXiv:9605213].

[22] I. Vitev and B. W. Zhang, "The Physics of Z^0/γ^*-tagged jets at the LHC", (2010) [arXiv:1006.2389].

[23] CMS Collaboration, "Determination of Jet Energy Calibration and Transverse Momentum Resolution in CMS", CMS-JME-10-011, CERN-PH-EP-2011-102,[arXiv:1107.4277].

[24] CMS Collaboration, "Study of Z boson production in PbPb collisions at nucleon-nucleon centre of mass energy = 2.76 TeV", Phys. Rev. Lett. 106 (2011) 212301,[arXiv:1102.5435].

[25] CMS Collaboration, "Studies of jet quenching using isolated-photon+jet correlations in PbPb and pp collisions at $\sqrt{s_{NN}} = 2.76$ TeV", Phys. Lett. B 718 (2013) 773.

[26] CMS Collaboration, "Nuclear modification factor of high transverse momentum jets in PbPb collisions at $\sqrt{s_{NN}}$ =2.76 TeV", Physics Analysis Summary, CMS-HIN-12-004 (2012).

[27] I. Vitev and B. W. Zhang, "A Theory of jet shapes and cross sections: From hadrons to nuclei", JHEP 0811 (2008) 093, [arXiv:0810.2807].

[28] B.A. Cole, G.G. Barnafoldi, P. Levai, G. Papp, G. Fai, "EMC effect and jet energy loss in relativistic deuteron-nucleus collisions", [arXiv:hep-ph/0702101].

[29] I. Vitev, "Non-Abelian energy loss in cold nuclear matter", Phys.Rev. C75 064906 (2007).

[30] J. J. Aubert, et al., "The ratio of the nucleon structure functions F2n for iron and deuterium", Phys. Lett. B123 (1983) 275.

[31] J. Gomez, et al., "Measurement of the A-dependence of deep inelastic electron scattering", Phys. Rev. D49 (1994) 4348.

[32] A. Majumder, "A comparative study of Jet-quenching Schemes", J. Phys. G 34 (2007) S377, [arXiv:nucl-th/0702066].

[33] N. Armesto, L. Cunqueiro and C. Salgado, "Monte Carlo for Jet Showers in the Medium", Nucl. Phys. A 830 (2009) 271, [arXiv:0907.4706].

[34] CMS Collaboration, "Modification of jet shapes in PbPb collisions at sqrt(s[NN]) = 2.76 TeV", [arXiv:1310.0878] (2013).

[35] CMS Collaboration, "Detailed Characterization of Jets in Heavy Ion Collisions Using Jet Shapes and Jet Fragmentation Functions", Physics Analysis Summary, CMS-HIN-12-013 (2012).

[36] Dokshitzer, Yuri L. and Kharzeev, D.E, "Heavy quark colorimetry of QCD matter", Phys. Lett. B519, (2001), 199.

[37] CMS Collaboration, "Suppression of non-prompt J/ψ, prompt J/ψ, and $Y(1S)$ in PbPb collisions at $\sqrt{s_{NN}}$ = 2.76 TeV", JHEP 1205 (2012) 063.

[38] CMS Collaboration, "Inclusive b-jet production in pp collisions at $\sqrt{s} = 7$ TeV", JHEP 1204 (2012) 084.

[39] CMS Collaboration, "Identification of b-quark jets with the CMS experiment". JINST 8 (2013) P04013, [arXiv:1211.4462].

[40] CMS Collaboration, "Evidence of b-jet quenching in PbPb collisions at $\sqrt{s_{NN}} = 2.76$ TeV", [arXiv:1312.4198] (2013).

[41] M. Ngyuen for the CMS Collaboration, "b-jet Identification in PbPb Collisions with CMS", Nucl.Phys. A904-905 (2013) 705c-708c, [arXiv:1211.5285].

[42] J. Robles for the CMS Collaboration, "Measurement of b-jet to inclusive jet ratio in PbPb and pp collisions at $\sqrt{s_{NN}} = 2.76$ TeV with the CMS detector ", Nucl.Phys. A904-905 (2013) 1011c-1014c.

Journal of Physics: Conference Series **589** (2015) 012018 doi:10.1088/1742-6596/589/1/012018

Strangeness production associated to a high-$p_{\rm T}$ particle in Pb–Pb collisions at $\sqrt{s_{\rm NN}} = 2.76$ TeV with ALICE

Xitzel Sánchez Castro for the ALICE Collaboration

IPHC, Université de Strasbourg, CNRS-IN2P3, 23 rue du loess, BP28, 67037 Strasbourg cedex 2, FR

E-mail: xsanchez@cern.ch

Abstract.
Particle production in heavy-ion collisions at RHIC and the LHC reveals an enhancement of the baryon-to-meson ratios with respect to pp collisions. The observed suppression of high-$p_{\rm T}$ particles raises the question of how the jet fragments may (re)distribute in ultra-relativistic heavy-ion collisions. With two-hadron correlations, we can study the particle composition in jets in order to get a better understanding of the aforementioned observables and the particle production mechanisms in the dense medium. In these proceedings, we present the status of the current analysis using two-hadron angular correlations for Pb–Pb collisions at $\sqrt{s_{\rm NN}} = 2.76$ TeV using the ALICE detector. We study the two different possibilities, considering the strange particle as either the trigger particle or the associated particle. Results on $\rm K^0_S$-$\rm h^{\pm}$, $(\Lambda + \overline{\Lambda})$-$\rm h^{\pm}$ and $\rm h^{\pm}$-$\rm K^0_S$ correlations are shown for central collisions.

1. Introduction

The main purpose of studying heavy-ion collisions at ultra-relativistic energies is to understand the phase transition of nuclear matter into a state of deconfined partons: the so called Quark-Gluon Plasma. At RHIC and LHC energies, a large enhancement of baryon-to-meson ratios is observed in the intermediate transverse momentum ($p_{\rm T}$) range (from 2 to 6 GeV/c) of central collisions, as compared to peripheral and to pp collisions [1, 2, 3]. The recent ALICE results in Pb–Pb collisions show this phenomenon, with the measurement of $\Lambda/\rm K^0_S$ ratio for different collision centralities [1]. An interplay between parton coalescence and parton fragmentation is suggested to explain the baryon-to-meson enhancement at intermediate $p_{\rm T}$ [4, 5]. Other models that include hydrodynamical radial flow in their description intend also to explain the baryon enhancement [6].

Studies of two-particle correlations, using strange hadrons as associated particles, can help to understand better whether the baryon enhancement is an effect from the particles produced in the thermalized medium, or whether the fragmented hadrons from the hard scattering between partons are contributing as well [7]. This hypothesis can be verified by separating particles produced together with a high-$p_{\rm T}$ particle and from the bulk. Consequently, we can estimate the baryon-to-meson ratio associated to each hadronization mechanism and study their relative contribution according the transverse momentum range.

Journal of Physics: Conference Series **589** (2015) 012018 doi:10.1088/1742-6596/589/1/012018

Additional information about the color-charge interaction of partons with the medium, via quark and gluon fragmentation, can be inferred on a statistical basis using the two-hadron angular correlation technique. Previous measurements have shown that gluon jets have a larger particle multiplicity with softer fragmented hadrons and are found to be broader than quark jets. It has also been shown that, compared with mesons, baryons are more likely to be produced in gluon jets [8].

In these proceedings we present the status of the ongoing analysis performed with strange particles using the two-particle correlation technique. The data used for the analysis were recorded for Pb–Pb collisions at $\sqrt{s_{\mathrm{NN}}} = 2.76$ TeV during the 2011 run using the ALICE detector at the LHC.

2. The ALICE experiment

A detailed description of the ALICE experiment can be found in [9]. It consists of a forward muon detector and a number of central detectors which fully cover a pseudorapidity range of $|\eta| < 0.9$. The central detectors are embedded in a 0.5 T magnetic solenoid with the field parallel to the beam direction (z-axis). The Inner Tracking System (ITS) and the Time Projection Chamber (TPC) are the main detectors for track and vertex finding, covering an acceptance of $|\eta| < 0.9$. Using these two subsystems, we reconstruct neutral kaons and Λ exploiting the topology of the following weak decays: $\mathrm{K^0_S} \to \pi^+\pi^-$, $\Lambda \to \mathrm{p}^+\pi^-$. The main advantage of working with these strange particles is that we can apply the same reconstruction technique for both $\mathrm{K^0_S}$ (meson) and Λ (baryon) over a wide momentum range from 0.4 GeV/c (0.6 GeV/c for Λ) to 12 GeV/c. The events were collected using an interaction trigger based on the information coming from the VZERO scintillator counters. Only events with coincident signal recorded in the VZERO-C and VZERO-A detectors were accepted. The centrality of the collision is determined with the VZERO detector and the Zero Degree Calorimeters (ZDCs) [10].

3. Correlations with strange particles

Using the most central Pb–Pb collisions, two different cases are considered for the correlations, the strange particle is used as: (i) the trigger particle (ii) the associated particle in the event. We select a particle with a high p_{T} (trigger particle) and the associated particles within a given p_{T} interval, both taken from the same event. The associated particles must have a transverse momentum lower than the trigger one ($p_{\mathrm{T},Assoc} < p_{\mathrm{T},Trig}$). The correlation is measured as a function of the distance in the azimuthal direction ($\Delta\varphi = \varphi_{Trig} - \varphi_{Assoc}$) and in pseudorapidity ($\Delta\eta = \eta_{Trig} - \eta_{Assoc}$) between the two particles. The obtained two-dimensional distribution is normalized to the total number of trigger particles (N_{Trig}). The signal distribution $S(\Delta\varphi, \Delta\eta) = 1/N_{Trig}\, \mathrm{d}^2N_{same}/\mathrm{d}\Delta\varphi\, \mathrm{d}\Delta\eta$ is corrected for the detector acceptance using mixed-event correlations ($B(\Delta\varphi, \Delta\eta) = \alpha\, \mathrm{d}^2N_{mixed}/\mathrm{d}\Delta\varphi\, \mathrm{d}\Delta\eta$). The mixed events are constructed by correlating a trigger particle in one event with associated particles in another event, with the condition that both events have similar z-vertex positions within the 2-cm-wide and a 5% centrality interval. The mixed-event distributions are scaled by a factor α that is chosen in such a way that $B(\Delta\varphi, \Delta\eta)$ is equal to unity in $\Delta\varphi \approx 0$ and $\Delta\eta \approx 0$.

The acceptance-corrected distribution is

$$\frac{1}{N_{Trig}} \frac{\mathrm{d}^2N_{Assoc}}{\mathrm{d}\Delta\varphi\, \mathrm{d}\Delta\eta} = \frac{S(\Delta\varphi, \Delta\eta)}{B(\Delta\varphi, \Delta\eta)}. \quad (1)$$

Further corrections, such as single track efficiency, contamination from decays of secondary particles and track merging/splitting, still need to be applied.

Figure 1 shows the acceptance-corrected distributions (illustrated in the range $|\Delta\eta| < 1.5$) with $\mathrm{K^0_S}$ and Λ (or $\overline{\Lambda}$) for trigger particles selected in the range $6 < p_{\mathrm{T}} < 15$ GeV/c in the

Journal of Physics: Conference Series **589** (2015) 012018 doi:10.1088/1742-6596/589/1/012018

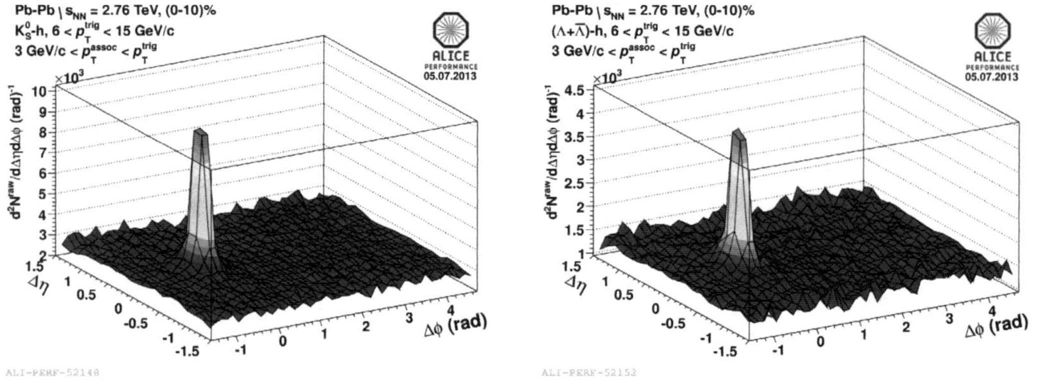

Figure 1. K^0_S-$h^{\pm}$ correlations (left plot) and $(\Lambda + \overline{\Lambda})$-$h^{\pm}$ correlations (right plot) in the 0-10% centrality interval. The distributions are acceptance-corrected only.

0-10% centrality interval. The K^0_S and Λ candidates are selected in the invariant mass range $0.48 < m_{\pi^+\pi^-} < 0.52$ GeV/c^2 and $1.108 < m_{p^+\pi^-} < 1.125$ GeV/c^2, respectively. Primary charged particles are selected as associated particles in the range 3 GeV/$c < p_{T,Assoc} < p_{T,Trig}$. The total number of analyzed events is about 30×10^6.

Further extraction of the signal is done by the method commonly known as η-gap [11], which assumes that the contribution from elliptic flow is $\Delta\eta$-independent. The bulk production is estimated in the range $1.0 < |\Delta\eta| < 1.8$, where no contribution from jets is expected, and it is subtracted from the near-side correlation in the acceptance range $|\Delta\eta| < 0.8$ to isolate the jet-like correlations.

At low p_T, the combinatorial background in the invariant mass distribution of K^0_S and Λ is significant, and therefore the correlation analysis requires special attention to remove the background. In Figure 2 we present an example of the $h^{\pm} - K^0_S$ angular correlations for events selected in 0-5% interval of the collision centrality and within $|\Delta\eta| < 1$. Distributions are neither corrected for the acceptance nor for the reconstruction efficiency yet. The trigger particles are unidentified charged particles in the range $5 < p_{T,Trig} < 10$ GeV/c. Neutral kaon

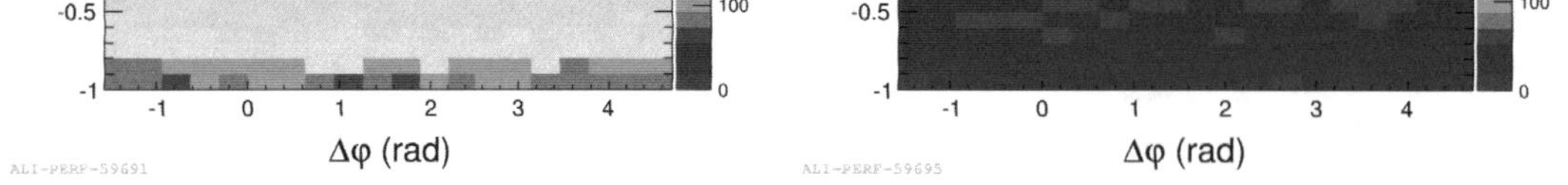

Figure 2. Correlations of K^0_S candidates with a charged trigger particle ($h^{\pm}$-K^0_S). Distributions are obtained by selecting the invariant mass region of the K^0_S candidates: $|m_{\pi^+\pi^-} - m_{K^0_S}| < 3\sigma$ (left plot signal) and $5\sigma < |m_{\pi^+\pi^-} - m_{K^0_S}| < 8\sigma$ (right plot background).

Journal of Physics: Conference Series **589** (2015) 012018 doi:10.1088/1742-6596/589/1/012018

candidates ($3.0 < p_{\mathrm{T},Assoc} < 3.5$ GeV/c) are selected in two different invariant mass region: $|m_{\pi^+\pi^-} - m_{\mathrm{K^0_S}}| < 3\sigma$ (signal region) and $5\sigma < |m_{\pi^+\pi^-} - m_{\mathrm{K^0_S}}| < 8\sigma$ (background region). It is noticeable that candidates in the background region present a correlation with the trigger particle. This feature makes the analysis challenging. A similar structure is observed for the case when the Λ baryon is used as the associated particle. Studies to obtain the ratio of the Λ and $\mathrm{K^0_S}$ yields associated to jets and from the bulk sample in the near-side region are ongoing, for the $p_{\mathrm{T},Assoc}$ interval 2-7 GeV/c and for more peripheral collisions. The origin of the baryon-to-meson enhancement could be understood by comparing the $\Lambda/\mathrm{K^0_S}$ in the bulk and in the jet with the inclusive analysis.

4. Summary

We have presented the current status of two-particle correlations using strange hadrons as trigger particles ($\mathrm{K^0_S}, \Lambda$) and as associated particles ($\mathrm{K^0_S}$) in central Pb–Pb collisions at $\sqrt{s_{\mathrm{NN}}} = 2.76$ TeV. These studies with identified hadrons may help to understand (i) the interaction of quark- and gluon-induced jets with the medium (using an identified hadron, $\mathrm{K^0_S}$ or Λ, as a trigger particle) and (ii) the origin of the baryon-to-meson enhancement at intermediate p_{T} by separating the contribution of the particles produced in the jet region and in the bulk region. The large sample of Pb–Pb collisions at $\sqrt{s_{\mathrm{NN}}} = 2.76$ TeV collected with the ALICE detector during the 2011 run at the LHC allows for both measurements in the near future.

References

[1] B. B. Abelev *et al.* [ALICE Collaboration] 2013 *Phys. Rev. Lett.* **111** 222301 [arXiv:1307.5530 [nucl-ex]]
[2] K. Adcox *et al.* [PHENIX Collaboration] 2004 *Phys. Rev.* C **69** 024904 [arXiv:0307010 [nucl-ex]]
[3] B. I. Abelev *et al.* [STAR Collaboration] 2006 *Phys. Rev. Lett.* **97** 152301 [arXiv:0606003 [nucl-ex]]
[4] R. J. Fries, V. Greco and P. Sorensen 2008 *Ann. Rev. Nucl. Part. Sci.* **58** 177 [arXiv:0807.4939 [nucl-th]]
[5] R. C. Hwa and C. B. Yang 2004 *Phys. Rev.* C **70** 024905 [arXiv:0401001 [nucl-th]]
[6] B. Abelev *et al.* [ALICE Collaboration] 2012 *Phys. Rev. Lett.* **109**, 252301 [arXiv:1208.1974 [hep-ex]]
[7] J. -X. Zuo [for the STAR Collaboration] 2008 *J. Phys.* G **35** 044027 [arXiv:0710.4203 [nucl-ex]]
[8] K. Ackerstaff *et al.* [OPAL Collaboration] 1999 *Eur. Phys. J.* C **8** 241 [arXiv:9805025 [hep-ex]]
[9] K. Aamodt *et al.* [ALICE Collaboration] 2008 *JINST* **3** S08002
[10] B. Abelev *et al.* [ALICE Collaboration] 2013 *Phys. Rev.* C **88** 044909 [arXiv:1301.4361 [nucl-ex]]
[11] A. Morsch [arXiv:1207.7187 [nucl-ex]]

Journal of Physics: Conference Series **589** (2015) 012019 doi:10.1088/1742-6596/589/1/012019

How do quarks and gluons lose energy in the QGP?

M. J. Tannenbaum[1]

Physics Department, Brookhaven National Laboratory, Upton, NY 11973-5000 USA

E-mail: mjt@bnl.gov

Abstract. RHIC introduced the method of hard scattering of partons as an in-situ probe of the the medium produced in A+A collisions. A suppression, $R_{\rm AA} \approx 0.2$ relative to binary-scaling, was discovered for π^0 production in the range $5 \leq p_T \leq 20$ GeV/c in central Au+Au collisions at $\sqrt{s_{NN}} = 200$ GeV, and surprisingly also for single-electrons from the decay of heavy quarks. Both these results have been confirmed in Pb+Pb collisions at the LHC at $\sqrt{s_{NN}} = 2.76$ TeV. Interestingly, in this p_T range the LHC results for pions nearly overlap the RHIC results. Thus, due to the flatter spectrum, the energy loss in the medium at LHC in this p_T range must be $\sim 40\%$ larger than at RHIC. Unique at the LHC are the beautiful measurements of the fractional transverse momentum imbalance $1 - \langle \hat{p}_{T_2}/\hat{p}_{T_1} \rangle$ of di-jets in Pb+Pb collisions. At the Utrecht meeting in 2011, I corrected for the fractional imbalance of di-jets with the same cuts in p-p collisions and showed that the relative fractional jet imbalance in Pb+Pb/p-p is $\approx 15\%$ for jets with $120 \leq \hat{p}_{T_1} \leq 360$ GeV/c. CMS later confirmed this much smaller imbalance compared to the same quantity derived from two-particle correlations of di-jet fragments at RHIC corresponding to jet $\hat{p}_T \approx 10-20$ GeV/c, which appear to show a much larger fractional jet imbalance $\approx 45\%$ in this lower $\hat{p}_T$ range. The variation of apparent energy loss in the medium as a function of both p_T and $\sqrt{s_{NN}}$ is striking and presents a challenge to both theory and experiment for improved understanding. There are many other such unresolved issues, for instance, the absence of evidence for a $\hat{q}$ effect, due to momentum transferred to the medium by outgoing partons, which would widen the away-side di-jet and di-hadron correlations in a similar fashion as the k_T-effect. Another issue well known from experiments at the CERN ISR, SpS and SpS collider is that parton-parton hard-collisions make negligible contribution to multiplicity or transverse energy production in p-p collisions—soft particles, with $p_T \leq 2$ GeV/c, predominate. Thus an apparent hard scattering component for A+A multiplicity distributions based on a popular formula, $dN_{\rm ch}^{AA}/d\eta = [(1-x)\,\langle N_{\rm part}\rangle\, dN_{\rm ch}^{pp}/d\eta/2 + x\,\langle N_{\rm coll}\rangle\, dN_{\rm ch}^{pp}/d\eta]$, seems to be an unphysical way to understand the deviation from $N_{\rm part}$ scaling. Based on recent p-p and d+A measurements, a more physical way is presented along with several other stimulating results and ideas from recent d+Au (p+Pb) measurements.

1. Introduction

There is a great textbook from 1950 titled "Nuclear Physics" from a course given by Enrico Fermi [1]. Chapter I of ten is "PROPERTIES OF NUCLEI" and Chapter II, "INTERACTION OF RADIATION WITH MATTER", which contains three sections: A) Energy Loss by Charged Particles; B. Scattering; C. Passage of Electromagnetic Radiation through Matter. Imagine that today we were going to write a textbook on QGP physics. After 13 years at RHIC and 2 years at LHC-IONS, we could do a fair job on Chapter I, "PROPERTIES OF THE QGP"; but for Chapter II "INTERACTION OF QUARKS AND GLUONS AND OTHER RADIATION WITH

1 Supported by the U.S. Department of Energy, Contract No. DE-AC02-98CH10886.

Journal of Physics: Conference Series **589** (2015) 012019 doi:10.1088/1742-6596/589/1/012019

THE QGP", I don't know. I have not seen evidence that provides convincing proof of any theory or model. I don't know any formula comparable to Bethe-Bloch for ionization loss or Bethe-Heitler for radiation. For instance, I would like to see something for quarks and gluons in the QGP like Fig. 1 for a muon in Cu. Of course the situation is much more complicated for the

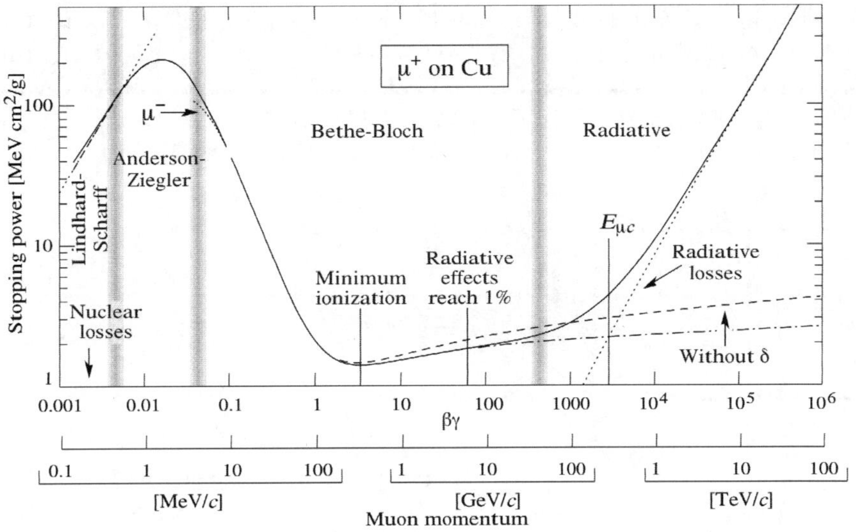

Figure 1. dE/dx of a μ^+ in Copper as a function of muon momentum [2]

QGP because it is not simply a block of material but an expanding and flowing hot medium.

There are many different theoretical studies of energy loss of a quark or gluon with their color charges fully exposed passing through a medium with a large density of similarly exposed color charges (i.e. a QGP). The approaches are different, but the one thing that they have in common [3] is the transport coefficient of a gluon in the medium, denoted $\hat{q}$, which is defined from the mean 4-momentum transfer^2/collision but is expressed as the mean 4-momentum transfer^2 per mean free path of a gluon in the medium. Thus the mean 4-momentum transfer^2 for a gluon traversing length L in the medium is, $\langle q^2(L)\rangle = \hat{q}\, L = \mu^2\, L/\lambda_{\rm mfp}$, where $\lambda_{\rm mfp}$ is the mean free path for a gluon interaction in the medium, and μ, the mean momentum transfer per collision, is the Debye screening mass acquired by gluons in the medium. In this, the original BDMPSZ formalism [4], the energy loss of an outgoing parton due to coherent gluon bremsstrahlung per unit length (x) of the medium, $-dE/dx$, takes the form [5]:

$$\frac{-dE}{dx} \simeq \alpha_s \left\langle q^2(L)\right\rangle = \alpha_s\, \hat{q}\, L = \alpha_s\, \mu^2\, L/\lambda_{\rm mfp} \quad , \tag{1}$$

so that the total energy loss in the medium goes like L^2 [6]. Also the accumulated transverse momentum^2, $\langle k_\perp^2\rangle$, for a gluon traversing a length L in the medium is well approximated by $\langle k_\perp^2\rangle \approx \langle q^2(L)\rangle = \hat{q}\, L$.

Experiments at RHIC were the first to use hard-scattering as an in-situ probe of gluons and quarks traversing the medium in Relativistic Heavy Ion (RHI) collisions. Many important and beautiful results have been obtained using single inclusive measurements and two-particle correlations. A simple estimate shows that the $\langle k_\perp^2\rangle \approx \hat{q}\, L$ should be observable at RHIC via the broadening of di-hadron azimuthal correlations. Assume that for a trigger particle with p_{T_t} the away-parton traverses slightly more than half the 14 fm diameter medium for central collisions of Au+Au, say 8 fm. With a $\hat{q} = 1$ GeV^2/fm [3], this would correspond to $\langle k_\perp^2\rangle = \hat{q}\, L = 8$

GeV^2, compared to the measured [7] $\langle k_T^2 \rangle = 8.0 \pm 0.2$ (GeV/c)^2 for di-hadrons in p-p collisions[2] with roughly the same p_{T_t} and $p_T^{\rm assoc}$. This should be visible as an azimuthal width $\sim \sqrt{2}$ larger in Au+Au than in p-p collisions at $\sqrt{s_{NN}}$ =200 GeV.

However, there is no direct evidence as yet for broadening of di-hadron or di-jet correlations from the effect of $\hat{q}$. The best measurement so far is an early measurement from the STAR collaboration [8] (Fig. 2) of di-hadron azimuthal correlations which shows that for a trigger particle with transverse momentum $8 < p_{T_t} < 15$ GeV/c, the conditional yield of associated charged hadrons with $p_T^{\rm assoc}$ in the away-side peak can be fit to a Gaussian with a width of $\sigma_{\Delta\phi} = 0.24 \pm 0.07$ for d+Au and 0.20 ± 0.02 (0.22 ± 0.02) for 20%-40% (0-5%) centrality Au+Au collisions. The result is not statistically significant for a difference in $\sigma_{\Delta\phi}$ for d+Au and Au+Au.

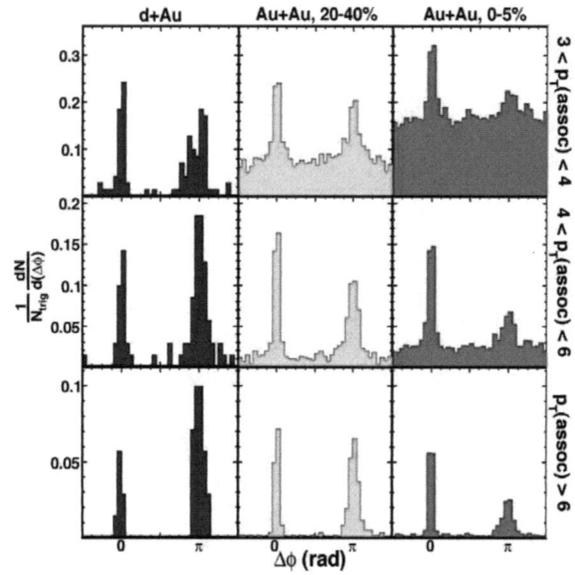

Figure 2. Di-hadron azimuthal correlations for charged hadron triggers with $8 < p_{T_t} < 15$ GeV/c and several values of $p_T^{\rm assoc}$ in minimum bias d+Au and 20%-40% (0-5%) centrality Au+Au collisions at $\sqrt{s_{NN}} = 200$ GeV [8]

There are many other such unresolved issues in RHI physics. In the following sections, I review some of the latest results at RHIC starting with some of the original results of hard-scattering and the discovery of "Jet Quenching" to place the latest results in context.

2. Jet Quenching in High p_T Physics at RHIC

The use of hard-scattering at RHIC as an in-situ probe of the medium produced in A+A collisions by the effect of the medium on outgoing hard-scattered partons has led to many important discoveries (and puzzles). The effect of the medium on outgoing hard-scattered partons produced

in the initial A+A collision is determined by comparison to measurements in p+A (or d+A) collisions, where no (or negligible) medium is produced.

[2] In both cases the azimuthal projection is only half the quoted $\langle k_T^2 \rangle$ or $\langle k_\perp^2 \rangle$

The discovery at RHIC [9] that hadrons with $p_T > 3$ GeV/c are suppressed in central Au+Au collisions by roughly a factor of 5 compared to point-like (binary) scaling from p-p collisions is arguably *the* major discovery in RHI physics. In order to verify that the suppression was due to the medium produced in Au+Au collisions and not an effect in the cold matter of an individual nucleus, measurements in d+Au collisions were performed in 2003 which were so definitive that all four experiments at RHIC had their results displayed on the front cover of Physical Review Letters (Fig. 3a). Figure 3b shows the Nuclear Modification factor $R_{\rm AA}$ for

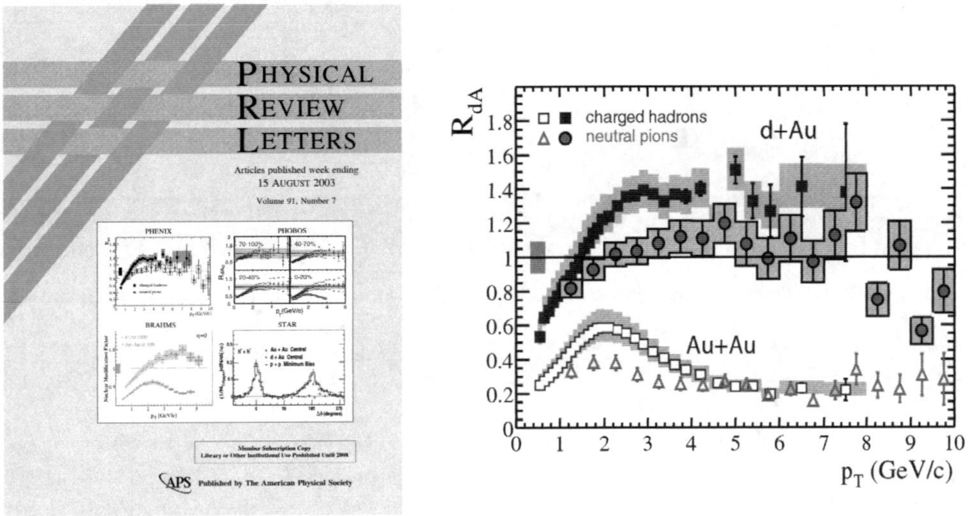

Figure 3. a) (left) Cover of Phys. Rev. Letters of 13 August 2003 displaying the results of all four RHIC experiments which showed no suppression in d+Au collisions. b) (right) PHENIX results from that volume [10] for d+Au and Au+Au collisions at $\sqrt{s_{NN}}$ =200 GeV.

non-identified charged hadrons ($h^{\pm}$) and π^0-mesons measured by PHENIX [10] in d+Au and Au+Au collisions at $\sqrt{s_{NN}}$ =200 GeV where $R_{\rm AA}(p_T)$ is the ratio of the yield of particles (h) for a given centrality Au+Au collision to the point-like-scaled p-p cross section,

$$R_{\rm AA}(p_T) = [d^2N^h_{AA}/dp_T dy N_{AA}]/[\langle T_{AA}\rangle\, d^2\sigma^h_{pp}/dp_T dy] \quad , \qquad (2)$$

where $\langle T_{AA}\rangle$ is the overlap integral of the nuclear thickness functions for that centrality. Two things are important to note about Fig. 3b: i) both the $h^{\pm}$ and π^0 are suppressed in Au+Au central collisions but not suppressed in d+Au collisions; ii) the $h^{\pm}$ and π^0 exhibit different values of $R_{\rm AA}$ in both d+Au and Au+Au collisions. This showed that supppression was a medium effect and that particle identification is crucial in these measurements because different particles behave differently.

The method of calculating the suppression of high p_T particles in Au+Au collisions at RHIC is shown in Fig. 4a for π^0 [11]. The hard-scattering in p-p collisions is indicated by the power law behavior p_T^{-n} for the invariant cross section, $Ed^3\sigma/dp^3$, with $n = 8.10\pm0.05$ for $p_T > 3$ GeV/c at $\sqrt{s_{NN}} = 200$ GeV. The Au+Au data at a given p_T can be characterized either as shifted lower in p_T by δp_T from the point-like scaled p-p data at $p'_T = p_T + \delta p_T$, or reduced in magnitude at the same p_T, i.e. suppressed. In Fig. 4b, the suppression of the many identified particles measured by PHENIX at RHIC over many years is presented as the Nuclear Modification Factor, $R_{\rm AA}$. These results show an interesting pattern. All particles are suppressed for $p_T \geq 2$ GeV/c, even the electrons ($e^{\pm}_{\rm HF}$) from c and b quark decay (but not the direct-γ). There is one notable exception—the protons are enhanced. This is called the baryon anomaly [12, 13]. The non-suppression of

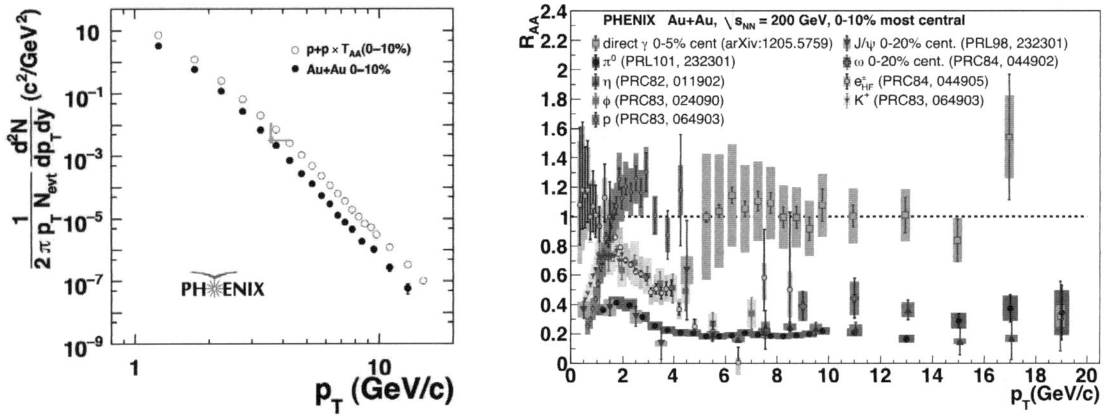

Figure 4. a) (left) Log-log plot of numerator and denominator of Eq. 2 as a function of p_T for π^0 at $\sqrt{s_{NN}}$ =200 GeV [11] b) PHENIX measurements of $R_{\rm AA}$ (p_T) of many identified particles in central Au+Au collisions at $\sqrt{s_{NN}}$ =200 GeV, with citations indicated.

the direct-γ in Au+Au collisions while the π^0, η and all the other mesons in Fig. 4 are suppressed shows that the suppression is an effect of the medium, the QGP, presumably due to the loss of energy by their parent quark or gluon with exposed color charge.

Direct-γ production at large p_T is one of most beautiful hard-scattering processes in QCD [14] as well as the best probe of the QGP because the γ ray participates directly in the predominant constituent reaction $g+q \to \gamma+q$ and emerges uscathed by the medium so that its momentum can be measured precisely. Also, since the scattered quark has equal and opposite p_T to the direct-γ, the p_T of the outgoing quark is known precisely at the production point. The simplicity and clarity of the constituent reaction stands in marked contrast to the difficulty of the experiment

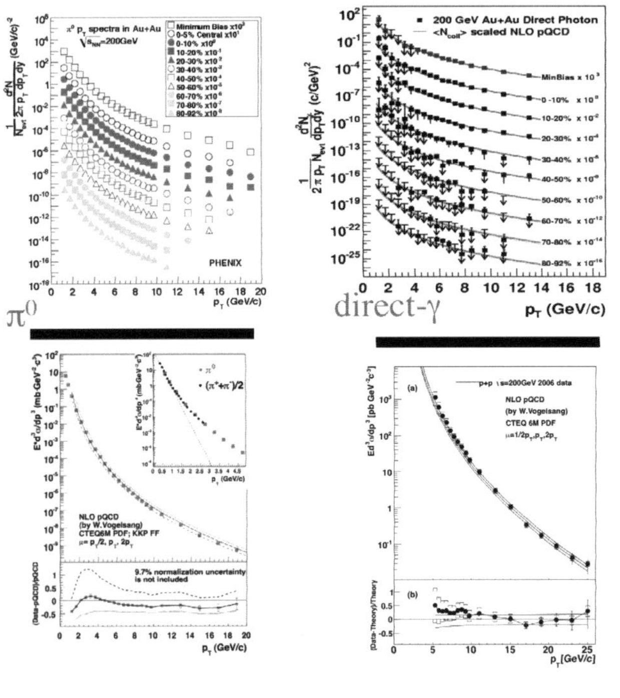

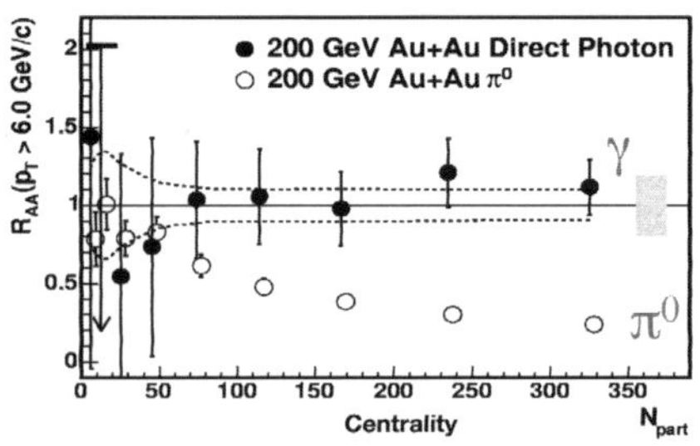

Figure 5. a) (left) PHENIX measurements of π^0 in Au+Au[15]/p-p[16] and direct-γ in Au+Au[17]/p-p[18]. b) $R_{\rm AA}$ (p_T) for π^0 and direct-γ as a function of centrality [17] from the measurements in (a).

Journal of Physics: Conference Series **589** (2015) 012019 doi:10.1088/1742-6596/589/1/012019

due to the overwhelming background of γ-rays from the γ+γ decay of π^0 and η mesons. However, with a properly designed experiment one can make four measurements, the π^0 and direct-γ p_T spectra in both p-p and A+A collisions which span many orders of magnitude (Fig. 5a), to obtain a precise comparison of $R_{\rm AA}$ for π^0 and direct-γ (Fig. 5b) [17]. The π^0 are more suppressed with increasing centrality while the direct-γ, with constant $R_{\rm AA} \approx 1$, are unaffected by the medium.

The dominance of the single constituent subprocess $g + q \to \gamma + q$ for direct-γ production allows an illustration of a fundamental property of QCD using the recent PHENIX measurements in p-p at $\sqrt{s}$ =200 GeV [18] combined with a collection of the world's measuements of direct-γ production in p-p and $\bar{\rm p}$+p collisions. Figure 6a shows an x_T scaling plot of all the measurements using the naive parton model value of $n_{\rm eff}$ = 4.0 [19, 20], which shows significant deviations compared to the beautiful x_T scaling (Fig. 6b) with index $n_{\rm eff}$ = 4.5. The difference of Figs. 6a and b illustrates the non-scaling of QCD due to the running coupling constant $\alpha_s(Q^2)$ and the evolution of the parton distribution functions [21].

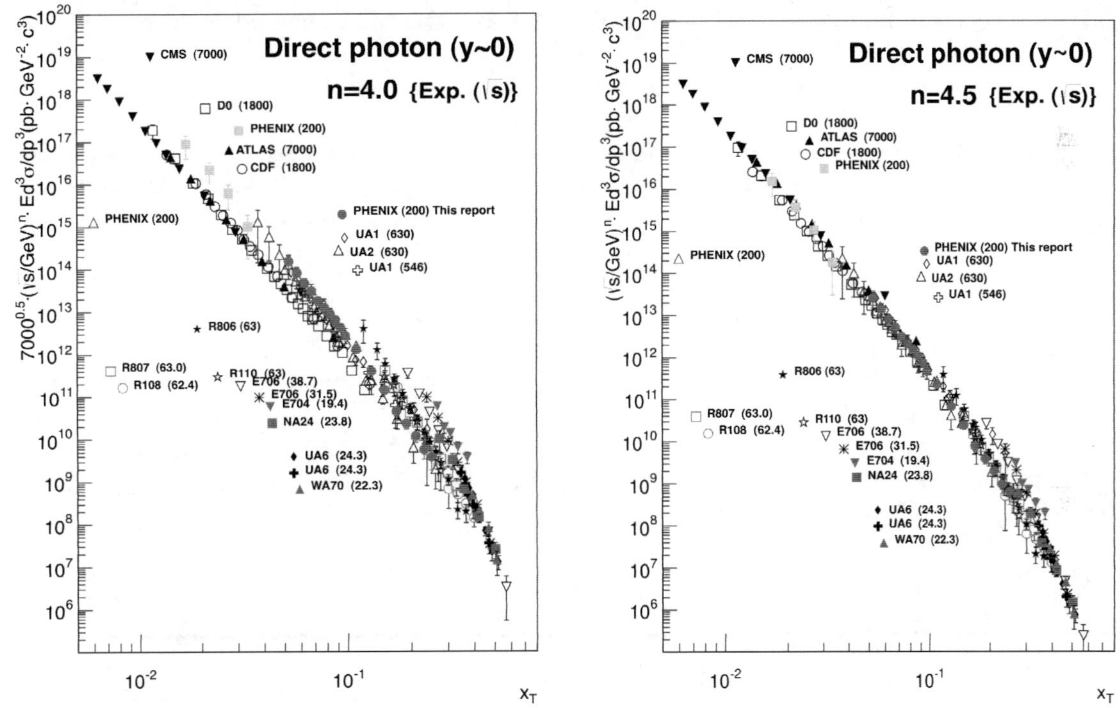

Figure 6. Direct-γ measurements plotted as $\sqrt{s}^{\,n_{\rm eff}} \times Ed^3\sigma/dp^3(x_T)$: a) (left) with $n_{\rm eff}$ = 4.0 b) (right) with $n_{\rm eff}$ = 4.5; where $x_T = 2p_T/\sqrt{s}$ [18]. The legend gives the experiment and $\sqrt{s}$.

For those who find the discussion in this section too sketchy, Jan Rak and I have published a book this year, with all this kind of information, which gives a full survey of "High p_T Physics in the Heavy Ion Era" over more than a century, from Rutherford (1911) to 2012, including many of the Pb+Pb results from the LHC [22].

3. New Results since Utrecht-2011

At the last meeting in this series that I attended, in Utrecht, the first results of the ALICE measurements of suppression of high p_T charged hadrons $h^{\pm}$ at $\sqrt{s_{NN}}$ =2.76 TeV at LHC were presented [23]. Despite more than a factor of 20 higher c.m. energy, the R_{AA} measurements by ALICE at LHC (Fig. 7a) [24] appear to be nearly identical to those from PHENIX at RHIC

Journal of Physics: Conference Series **589** (2015) 012019 doi:10.1088/1742-6596/589/1/012019

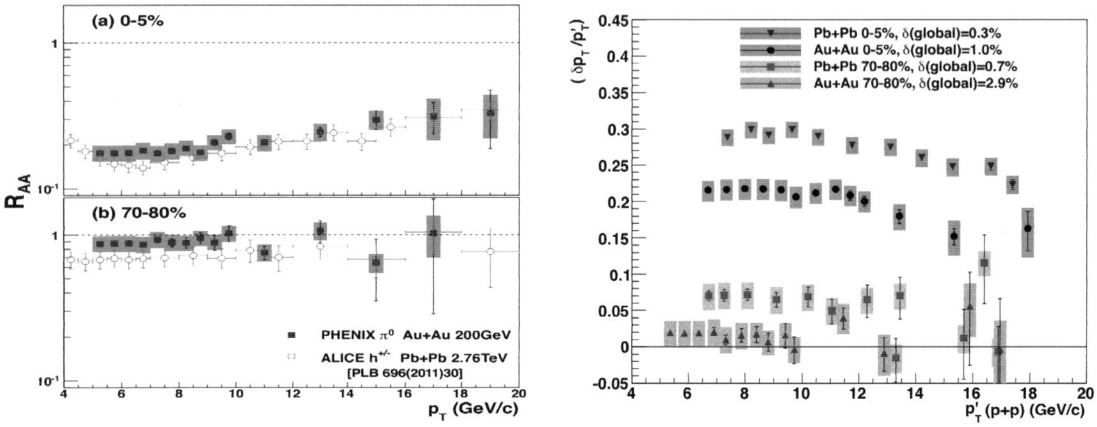

Figure 7. a) (left) R_{AA} of π^0 in $\sqrt{s_{NN}} = 200$ GeV central (0-5%) and peripheral (70-80%) Au+Au collisions [24] at RHIC compared to non-identified charged hadron ($h^{\pm}$) R_{AA} in $\sqrt{s_{NN}} = 2.76$ TeV Pb+Pb collisions at LHC. b) (right) Fractional shift of p_T spectrum $\delta p_T/p'_T$ vs. p'_T (p-p) calculated by PHENIX [24] for RHIC and LHC.

for $5 \leq p_T \leq 20$ GeV/c, except that for the LHC data, with better statistics, the upward trend of $R_{AA}(p_T)$ over the whole interval is significant (although the more recent PHENIX data in Fig. 7a [24] show a small but significant upward trend). However, what is most important to realize is that the exponent of the power-law at LHC ($n \approx 6$) is flatter than at RHIC ($n \approx 8$), so that to get the same value for R_{AA} at LHC as at RHIC, a $\sim 40\%$ larger shift $\delta p_T/p'_T$ in the spectrum from p-p to A+A is required. This led us more recently to compare the data in terms of $\delta p_T/p'_T$ (Fig. 7b) which clearly shows the larger shift for central collisions at LHC. The larger shift in the p_T spectrum likely indicates $\sim 40\%$ larger fractional energy loss at LHC than at RHIC in this p_T range due to the probably hotter and denser medium. These measurements can be combined with the previous measurements at RHIC for $\sqrt{s_{NN}} = 39$ and 62.4 GeV [25] (Fig. 8) to reveal a systematic increase of $\delta p_T/p'_T$ in central A+A collisions at $p'_T = 7$ GeV/c, going from 5% to 30% over the c.m. energy range $\sqrt{s_{NN}} = 39$ GeV to 2.76 TeV.

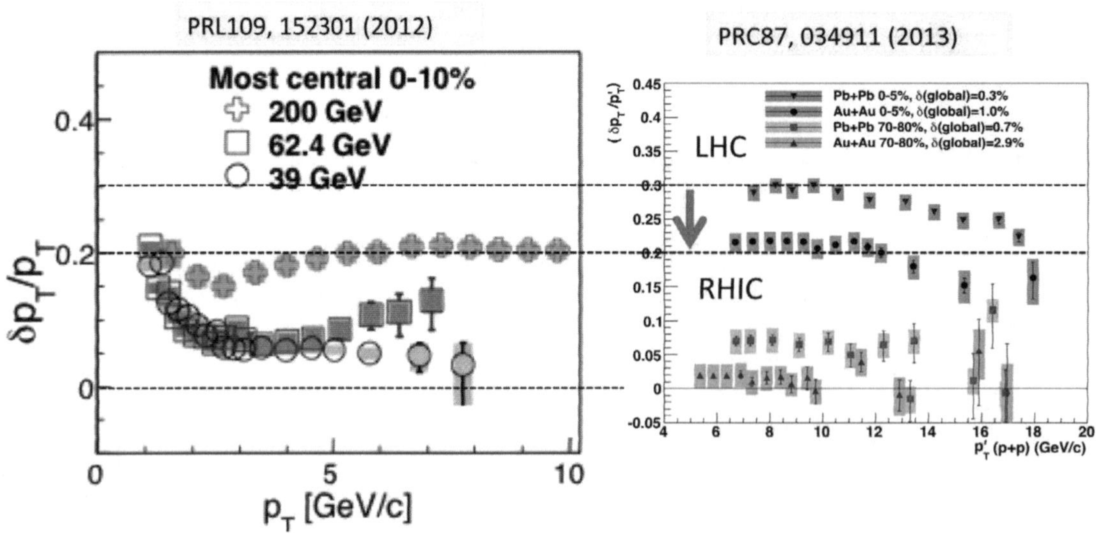

Figure 8. Fractional shift of p_T spectrum $\delta p_T/p'_T$ in central A+A collisions from $\sqrt{s_{NN}} =$ 39 GeV to 2.76 TeV

Journal of Physics: Conference Series **589** (2015) 012019 doi:10.1088/1742-6596/589/1/012019

At Utrecht, I also discussed another unexpected discovery at RHIC [7], that the distribution of particles, with p_{T_a}, opposite in azimuth to a trigger particle, e.g. a π^0 with large p_{T_t}, which is itself the fragment of a jet, does not measure the fragmentation function of the jet opposite in azimuth to the trigger, but, instead, measures the ratio of $\hat{p}_{T_a}$, of the away-parton, to $\hat{p}_{T_t}$, of the trigger-parton, and depends only on the same power n as the invariant single particle spectrum:

$$dP/dx_E|_{p_{T_t}} \approx N\,(n-1)/[\,\hat{x}_h(1+x_E/\hat{x}_h)^n] \quad . \tag{3}$$

This equation gives a simple relationship between the ratio, $x_E \approx p_{T_a}/p_{T_t}$, of the transverse momenta of the away-side particle to the trigger particle, and the ratio of the transverse momenta of the away-jet to the trigger-jet, $\hat{x}_h = \hat{p}_{T_a}/\hat{p}_{T_t}$. PHENIX measurements [26] of the x_E distributions of π^0-h correlations in p-p and Au+Au collisions at $\sqrt{s_{NN}} = 200$ GeV were fit to Eq. 3 (Fig. 9a,b) [27]. The results for the fitted parameters are shown on the figures. In

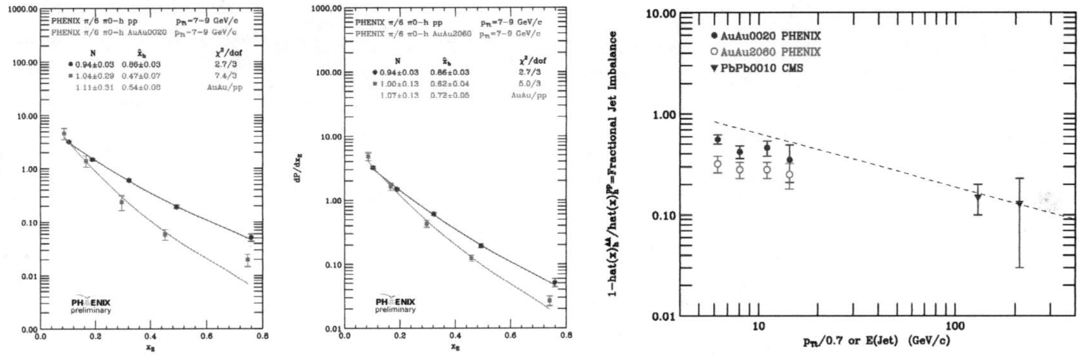

Figure 9. (left) x_E distributions at RHIC [27] from p-p (blue circles) and AuAu (red squares) collisions for $p_{T_t} = 7-9$ GeV/c, together with fits to Eq. 3 with parameters indicated: p-p (solid blue line) $N^{pp} = 0.94 \pm 0.03$, $\hat{x}_h^{pp} = 0.86 \pm 0.03$. For AuAu (solid red line), the ratios of the fitted parameters for AuAu/pp are also given: a) 00-20% centrality, $N^{AA} = 1.04 \pm 0.29$, $\hat{x}_h^{AA} = 0.47 \pm 0.07$, $\hat{x}_h^{AA}/\hat{x}_h^{pp} = 0.54 \pm 0.08$; b) 20–60% centrality $N^{AA} = 1.00 \pm 0.13$, $\hat{x}_h^{AA} = 0.62 \pm 0.04$, $\hat{x}_h^{AA}/\hat{x}_h^{pp} = 0.72 \pm 0.05$. c) (right) Fractional jet imbalance [27], $1 - \hat{x}_h^{AA}/\hat{x}_h^{pp}$, for the RHIC data from (a) and (b), and CMS data [28] as estimated by MJT [27]. The dashed line is an estimate [27] of the Fractional Jet Imbalance vs. E(Jet) at $\sqrt{s_{NN}}$ =2.76 TeV.

general the values of $\hat{x}_h^{pp}$ do not equal 1 but vary between $0.8 < \hat{x}_h^{pp} < 1.0$ due to k_T smearing and the range of x_E covered. In order to take account of the imbalance ($\hat{x}_h^{pp} < 1$) observed in the p-p data, the ratio $\hat{x}_h^{AA}/\hat{x}_h^{pp}$ is taken as the measure of the energy of the away jet relative to the trigger jet in A+A compared to p-p collisions.

The fractional jet imbalance was also measured directly with reconstructed di-jets by the CMS collaboration at the LHC in Pb+Pb central collisions at $\sqrt{s_{\rm NN}} = 2.76$ TeV (Fig. 10) [28]. At Utrecht, I calculated the same ratio $\hat{x}_h^{AA}/\hat{x}_h^{pp}$ for the CMS data to correct for the large effect in p-p collisions [27] and the results were compared to PHENIX as shown in Fig. 9c. The dashed line shows my estimate of the Fractional Jet Imbalance vs. E(Jet) at $\sqrt{s_{NN}}$ =2.76 TeV and includes the 40% increase in $\delta p_T/p'_T$ from Fig. 7b. New results in 2012 by CMS (Fig. 11) [29] significantly extend and improve their previous measurement and confirm my correction [27].

The large difference in fractional jet imbalance between RHIC and LHC c.m. energies could be due to the difference in jet $\hat{p}_{T_t}$ between RHIC ($\sim$ 20 GeV/c) and LHC ($\sim$ 200 GeV/c), a difference in the properties of the medium, the difference in n for the different $\sqrt{s_{NN}}$, or a problem with Eq. 3 which has not been verified by direct comparison to di-jets. In any case the strong $\hat{p}_T$ dependence of the fractional jet imbalance (apparent energy loss of a parton) also

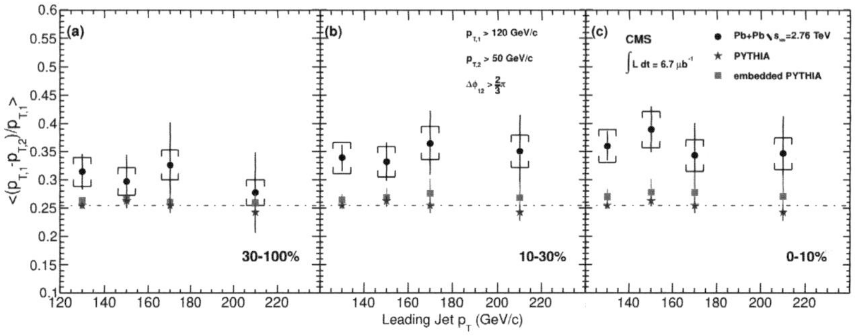

Figure 10. CMS measurement [28] of $\langle 1 - \hat{p}_{t,2}/\hat{p}_{t,1}\rangle$, the fractional jet imbalance, for 3 centralities in Pb+Pb collisions at $\sqrt{s_{NN}} = 2.76$ TeV, compared to PYTHIA simulations for p-p collisions.

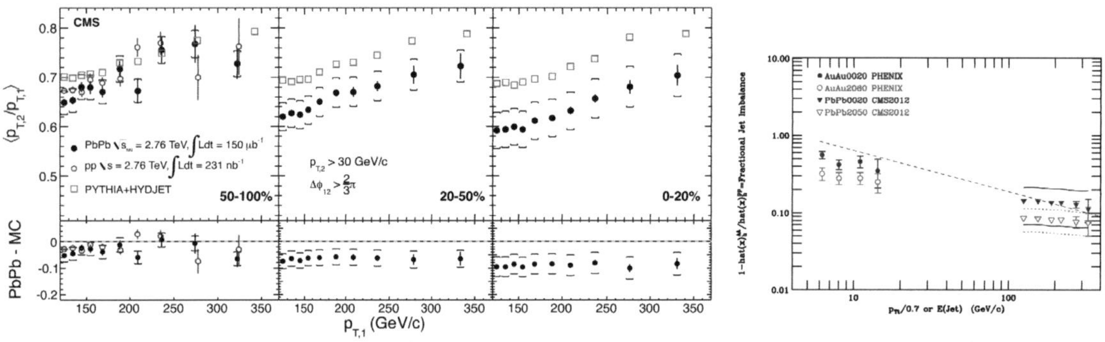

Figure 11. a) (left 3 panels) CMS [29] measurements of average di-jet transverse momentum ratio, $\hat{x}_h = \hat{p}_{T,2}/\hat{p}_{T,1}$, as a function of leading jet $\hat{p}_{T,1}$ at $\sqrt{s_{NN}} = 2.76$ TeV in p-p collisions and for 3 centralities in Pb+Pb collisions, as well as simulated p-p di-jets embedded in heavy ion events. b) (right) Fractional jet imbalance [27], $1 - \hat{x}_h^{AA}/\hat{x}_h^{pp}$, for the RHIC data from Fig. 9 with CMS measurement from (a). The solid (dotted) lines represent the systematic uncertainty of the CMS 0-20% (20-50%) results. The dashed line is the estimate of the Fractional Jet Imbalance vs. E(Jet) at $\sqrt{s_{NN}}$ =2.76 TeV from Fig. 9c.

seems to disfavor purely radiative energy-loss in the QGP [4] and indicates that the details of energy loss in a QGP remain to be understood. Future measurements at both RHIC and LHC will need to sort out these issues by extending di-jet and two-particle correlation measurements to overlapping regions of $\hat{p}_T$.

In distinction to the case of di-hadron correlations (where both hadrons are fragments of jets) which do not measure the fragmentation function, direct-γ-hadron correlations do measure the fragmentation function of the jet from the away parton (mostlikely a u quark [22]). PHENIX has shown that this is true for p-p collisions Fig. 12a [30]. This year, improved measurements by PHENIX [31] in both p-p and Au+Au collisions (Fig. 12b) now indicate a significant modification of the fragmentation function in Au+Au (0-40%) central collisions compared to p-p , with an enhancement at low $z_T = p_T^h/p_T^\gamma$ (large $\xi = -\ln z_T$) and a suppression at large z_T (small ξ) which is more clearly seen as $I_{AA}(\xi)$, the ratio of the fragmentation functions in Au+Au/pp (Fig. 12c). As shown in Fig. 12c, restricting the away-side azimuthal range reduces the large $\xi > 0.9$ ($p_T^h \lesssim 3$ GeV/c) enhancement but leaves the suppression at small $\xi < 0.9$ relatively unchanged, which shows that the large ξ enhancement is predominantly at large angles, similar to the effect observed by CMS with actual jets. [32].

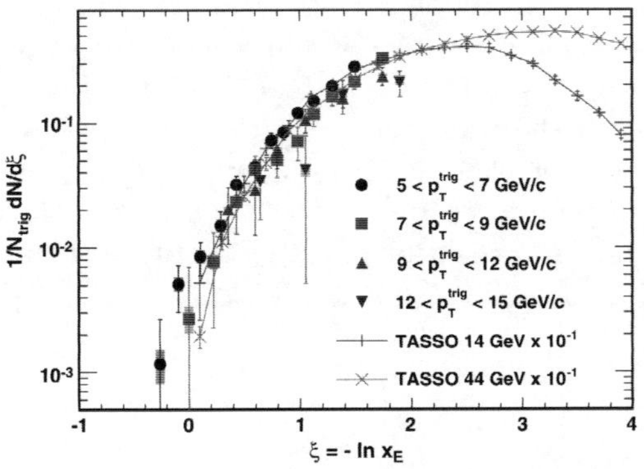

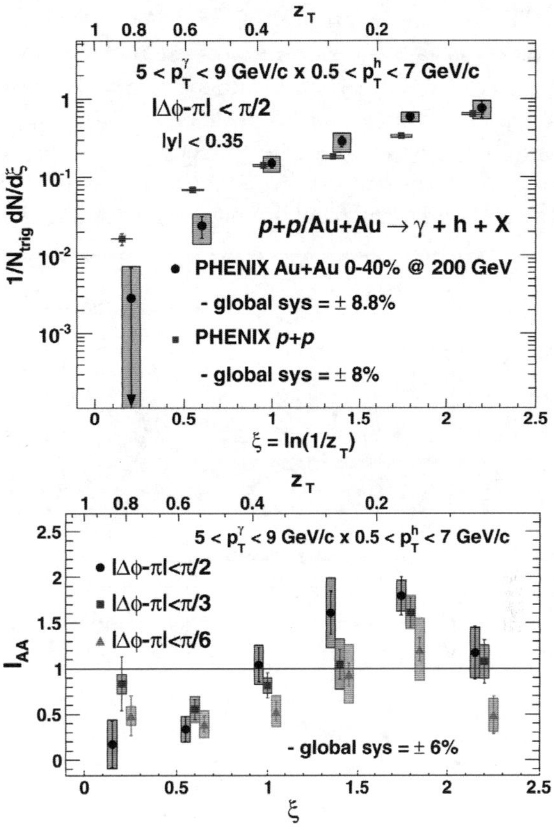

Figure 12. a) (left) PHENIX measurement [30] of direct-γ-h correlations in p-p collisions at $\sqrt{s}$ =200 GeV in the variable $\xi = -\ln x_E \approx -\ln(p_T^h/p_T^\gamma)$ compared to fragmentation functions measured in e^+e^- collisions at $\sqrt{s}$ =14 and 44 GeV by TASSO [30]. b)(right)-(top) ξ distributions of direct-γ-h correlations in Au+Au and p-p collisions at $\sqrt{s}$ =200 GeV. c) (right)-(bottom) Ratio of the Au+Au/p-p distributions, $I_{AA}(\xi)$ when the away side azimuthal range is restricted as indicated [31]

The principal difficulty in observing the broadening of di-jet or di-hadron azimuthal correlations by the transport coefficient $\hat{q}$ of the QGP stems from artifacts with names such as "Mach Cone", "Ridge", "Head and Shoulders" (Fig. 13). Perhaps one should have been suspicious because the artifact was observed in correlations triggered by both high (Fig 13a) and low (Fig 13b) p_T particles. These artifacts are now known to be due to odd harmonics

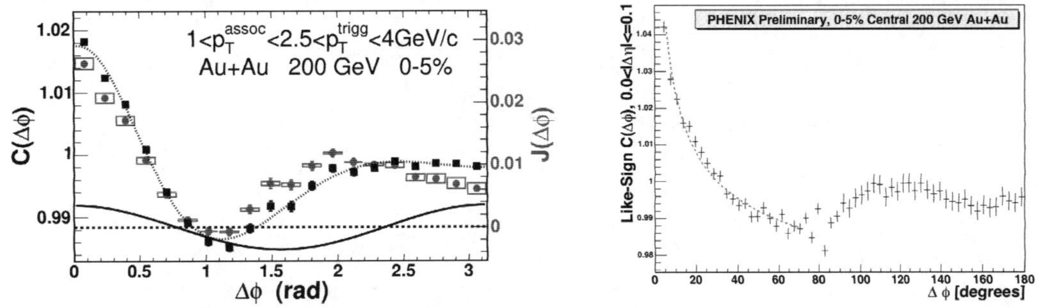

Figure 13. a) (left) Azimuthal correlation $C(\Delta\phi)$ (■) of $h^{\pm}$ with $1 \leq p_{T_a} \leq 2.5$ GeV/c from a "high p_T" trigger $h^{\pm}$ with $2.5 \leq p_{T_t} \leq 4$ GeV/c in Au+Au central (0-5%) collisions at $\sqrt{s_{NN}} = 200$ GeV [33]. The solid line is the correction for v_2 which is subtracted to give the jet correlation function $J(\Delta\phi)$ (●); b)(right) $C(\Delta\phi)$ for like sign $h^{\pm}$-pairs with $0.2 < p_{T_1}, p_{T_2} < 0.4$ GeV/c for both particles [34].

Journal of Physics: Conference Series **589** (2015) 012019 doi:10.1088/1742-6596/589/1/012019

of collective flow, notably v_3, which were thought to be forbidden by the assumed azimuthal symmetry $\phi \to \phi+\pi$ of the almond shaped participant region of an A+A collision but were shown (only recently [35]) to be produced from event-by-event fluctuations in the collision geometry.

Of course, understanding that the extra "bumps" in the correlation function are due to odd harmonics still requires one to know the values of these harmonics in order to subtract them. This is still the largest systematic uncertainty in attempts to observe the $\hat{q}$-broadening, for instance in a recent measurement by STAR of jet-hadron correlations [36] (Fig. 14). When the

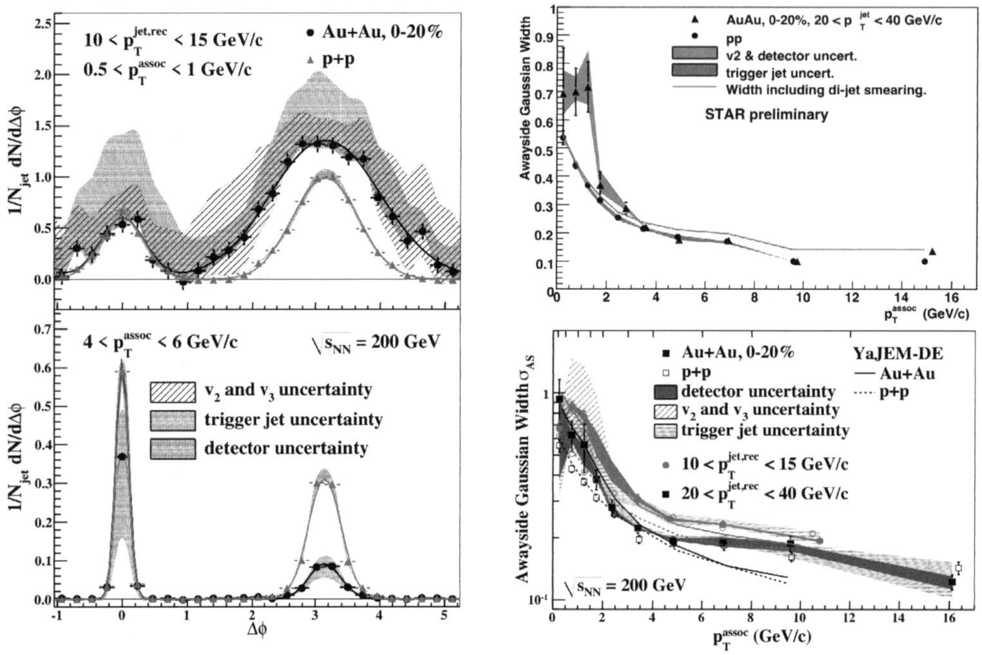

Figure 14. a) (left) Azimuthal correlation $J(\Delta\phi)$ with systematic uncertainties shown [36]. b)(right)-(top) Preliminary result [37] for Awayside rms width, $\sigma_{\rm AS}$, as a function of p_{T_a}. c)(right)-(bottom) Final result [36] for $\sigma_{\rm AS}$, as a function of p_{T_a}.

full systematic uncertainties, including those on v_2 and v_3 (Fig. 14a), are taken into account, the result for the medium induced broadening of the away-side widths, $\sigma_{\rm AS}$, in Au+Au relative to p-p which looked significant in the preliminary results (Fig. 14b) [37] become only "suggestive of medium-induced broadening [36]" in the final result (Fig. 14c) because "they are highly dependent on the shape of the subtracted background [36]", notably the v_2 and v_3 of the trigger jets.

One way out of this dilemma, at least for di-hadron correlations, has been pointed out by Roy Lacey and his group [38] who proposed "acoustic scaling" of the harmonics in analogy with attenuation of pressure driven sound waves in a viscous medium. They observe that viscous corrections damp the eccentricity-driven flow harmonics, v_n with increasing n but have minimal effect on the p_T dependence because of the small values of η/s. The proposed "acoustic scaling" that $v_n/(v_2)^{n-2}$ is independent of p_T was verified, at least for v_3 and v_4 (Fig. 15) [38], using previously published PHENIX measurements [39]. This opens the way to reduce the systematic uncertainty in subtracting the harmonic background to di-hadron correlations: once $v_2(p_T)$ is known or determined at any value of p_T and centrality, the acoustic scaling can be used to constrain the higher harmonics.

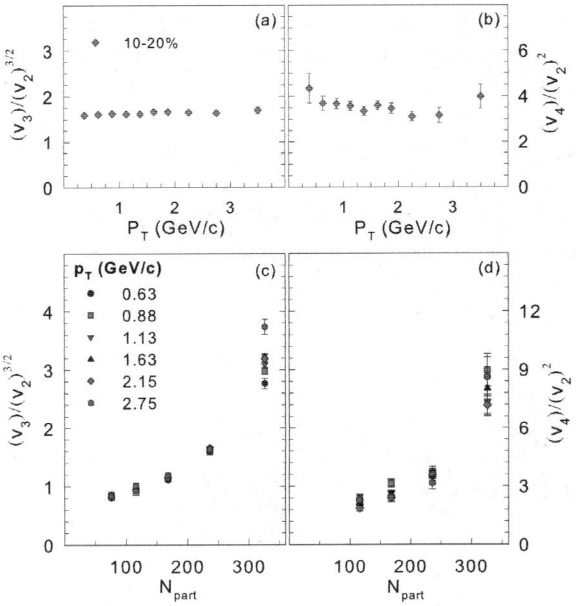

Figure 15. a) $v_3/(v_2)^{3/2}$ vs. p_T for 10-20% central Au+Au collisions at $\sqrt{s_{NN}}$ =200 GeV; b) same for v_4/v_2^2; c) $v_3/(v_2)^{3/2}$ vs centrality ($N_{\rm part}$) for several p_T cuts; d) same for v_4/v_2^2. Plot is from Ref. [38] using data from Ref. [39].

Another major topic this year was the p+Pb run at LHC which spurred new or improved d+Au results from RHIC. There is both clarity and confusion in the d+Au results. First, the clarity. Switching momentarily to soft multi-particle physics, new PHENIX measurements of E_T distributions at mid-rapidity in p-p, d+Au and Au+Au collisions at $\sqrt{s_{NN}}$ =200 GeV verify that starting from the p-p distribution, the Au+Au distribution can be calculated by taking the fundamental elements of particle production as the number of constituent-quark participants, as proposed in 2003 [40] (NQP model). For symmetric systems such as Au+Au the NQP model is equivalent to the Additive Quark Model (AQM) from 1982 [41] in which the fundamental element is a color string streched between constituent quark participants in the projectile and target with the additional restriction that only one color-string can be attached to a quark-participant. For asymmetric systems such as d+Au, the models differ because the number of color-strings is proportional only to the number of quark-participants in the smaller nucleus. In

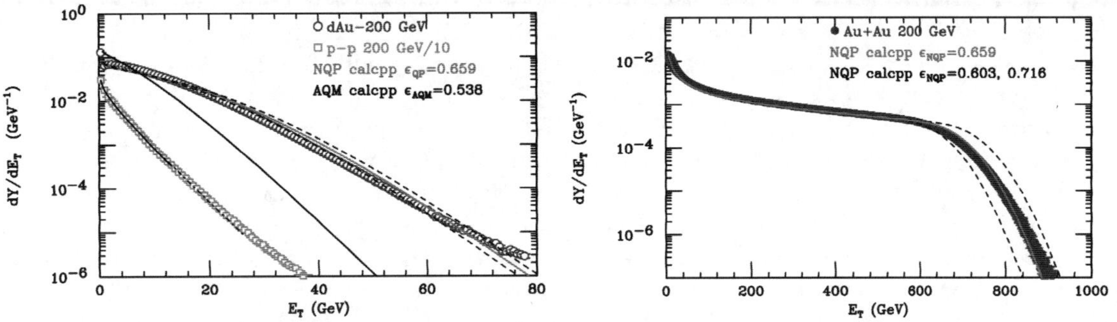

Figure 16. a) (left) PHENIX measurement [42] of mid-rapidity E_T distributions for p-p (□) and d+Au (○) at $\sqrt{s_{NN}}$ =200 GeV with calculations of the d+Au spectrum from fits to the p-p data (lines) based on the AQM (color-strings) and the number of constituent-quark participants (NQP). b) (right) Au+Au E_T distribution and NQP calculation. Systematic uncertainties are shown by dashed lines in both (a) and (b).

Journal of Physics: Conference Series **589** (2015) 012019 doi:10.1088/1742-6596/589/1/012019

Fig. 16 [42] the calculations which are based only on the measured p-p E_T distribution and Monte Carlo Glauber calculations of the nuclear geometry for the various cases show that the NQP calculation agrees with both the d+Au (Fig. 16a) and Au+Au (Fig. 16b) E_T measurements, but that the AQM exhibits a factor of 1.7 less E_T emission than observed in d+Au (Fig. 16a) due to the restriction on the number of effective constituent quark participants in the larger nucleus. In the Glauber calculation, the positions of the three constituent-quarks are generated about the position of each nucleon according to the measured charge distribution of the proton, which gives a physical basis for "proton size fluctuations" in nuclear geometry calculations.

On the hard-scattering front, new results for identified charged hadrons in d+Au (Fig. 17a) [43] may give a clue for understanding the baryon anomaly. In all centrality classes, the

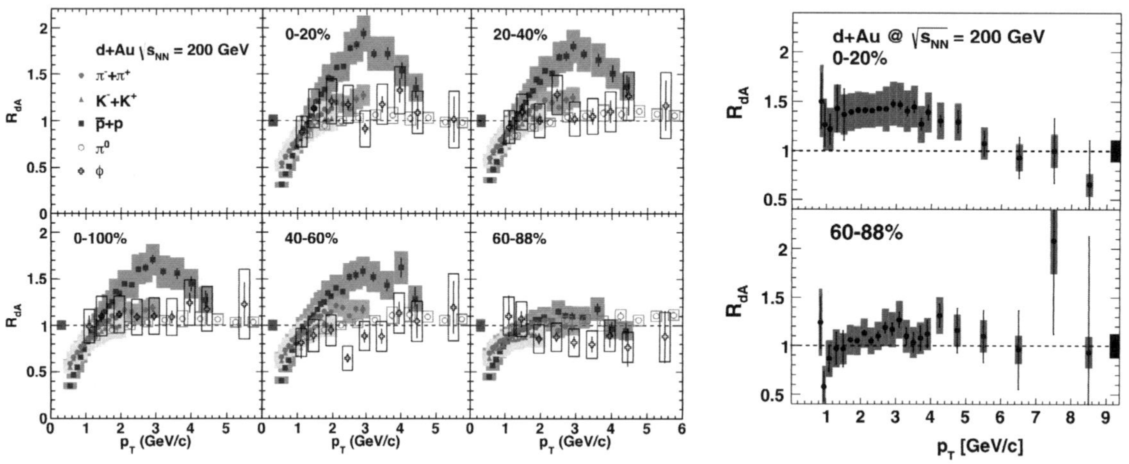

Figure 17. a) (left) PHENIX measurement [43] of R_{dA} of identified particles as a function of p_T and centrality in d+Au collisions at $\sqrt{s_{NN}}$ =200 GeV. b) (right) PHENIX measurement [44] of R_{dA} for direct-single $e^{\pm}$ from Heavy Flavor (c and b quark) decay for peripheral (60-88%) and central (0-20%) d+Au collisions at $\sqrt{s_{NN}}$ =200 GeV.

mesons, π^0, $\pi^{\pm}$, $K^{\pm}$ and ϕ show point-like behavior, $R_{\mathrm{dA}} = 1$, over the range $2 \leq p_T \leq 6$ GeV/c, while, apart from the most peripheral bin, the protons exhibit a huge enhancement (Cronin effect [45]) for $2 \leq p_T \leq 4$ GeV/c similar to the baryon anomaly in Au+Au (recall Fig. 4). In fact, as originally shown in a previous PHENIX measurement [46] and confirmed with the new more precise measurements [43], the proton enhancement in d+Au is generally considerably larger than in Au+Au. The new observation makes it clear that a common explanation of the d+Au and Au+Au baryon enhancements for $2 \leq p_T \lesssim 4 - 6$ GeV/c is needed.

It has been popular, though not necessarily correct [47, 48], to attribute the baryon anomaly to recombination of soft partons rather than fragmentation of hard-scattered partons [49, 50] because the suppression has been observed to be related to a "baryon vs. meson dynamic, as opposed to a simple mass-dependence [43]". However a new d+Au measurement of single $e^{\pm}_{\mathrm{HF}}$ from c and b heavy quarks, mostlikely from the decay of D-mesons with mass twice that of the proton, appears to create another layer of doubt to the recombination argument (Fig. 17b)[44]. In Au+Au (recall Fig. 4b), the $e^{\pm}_{\mathrm{HF}}$ are suppressed like the (K^+) mesons in the range $2 \leq p_T \leq 6$ GeV/c; but in d+Au (Fig. 17b) they show a Cronin enhancement like the baryons, not like the mesons, in central (0-20%) collisions. Thus the recombination issue is coupled to understanding the Cronin effect [45] which suffers from the problem that in the nearly 40 years since its discovery, the exact cause of the Cronin effect still remains a mystery.

Another confusing cold nuclear matter (CNM) effect concerns the definition of centrality in d+Au collisions. This is visible mostly in peripheral collisions for $p_T > 10$ GeV/c. In Fig. 17, the $R_{\rm dA}$ for peripheral (60-88%) collisions for $2 \leq p_T \leq 6$ GeV/c looked reasonable, i.e. $R_{\rm dA} \approx 1$ for all the particles, while in Fig. 18a [51], for higher p_T π^0, η mesons as well as jets , there appears to be a problem or something exciting according to one's taste. The peripheral collisions show a

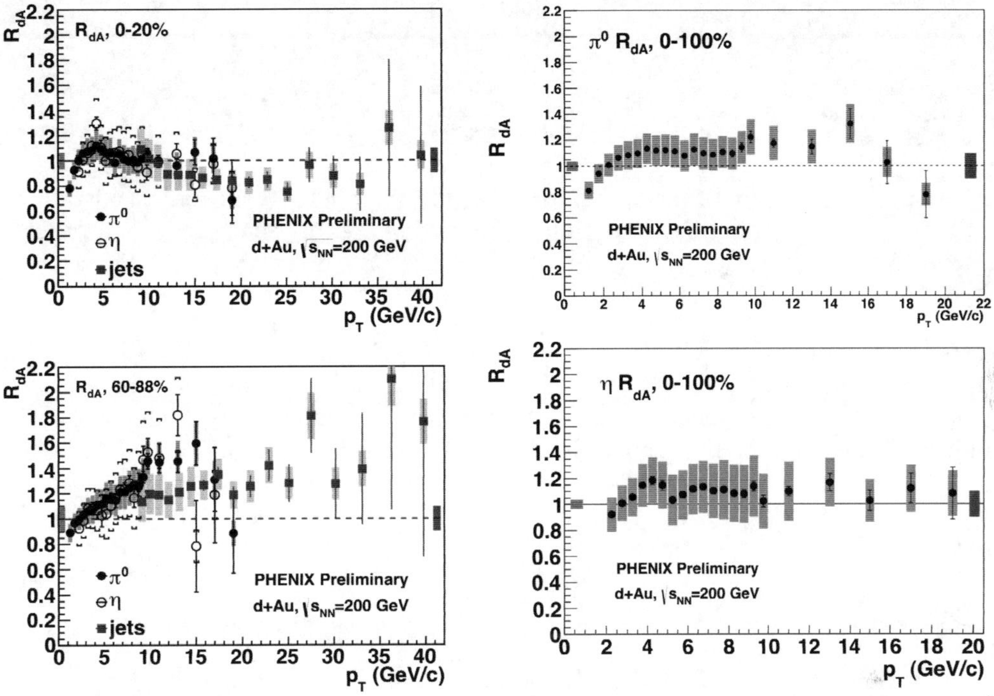

Figure 18. a) (left) $R_{\rm dA}$ for π^0, η and jets as a function of p_T for central (0-20%) and peripheral (60-88%) d+Au collisions at $\sqrt{s_{NN}}$ =200 GeV [51]. b)(right) $R_{\rm dA}$ for minimum bias π^0 and η [51].

huge enhancement (which doesn't make sense to the author because true peripheral collisions are largely single nucleon-nucleon collisions) while the central collisions don't look unreasonable. Personally, I think that we should only use minimum bias p+A collisions to understand the nuclear-size dependence of any CNM effect because the hard-scattering cross section in p+A collisions is simply A times the p-p cross section (modulo some possible isotopic spin effects, which are small). Glauber calculations and matching to some measure of centrality are not necessary. Also the comparison with parton distribution functions measured in e+A collisions is straightforward because centrality has generally not been measured in electron scattering and is likely not measurable. In fact, the minimum bias $R_{\rm dA}$ for π^0 and η from the same measurement [51] (Fig. 18b) look reasonable; and the minimum bias jet $R_{\rm dA}$ looks similar to the minimum bias π^0 $R_{\rm dA}$ rather than to the central collision jet $R_{\rm dA}$ [52]. A p+A run with several A is planned at RHIC for 2015-16 in addition to runs with ^{3}He+Au and possibly more d+Au.

For PHENIX, the future plans are built around a new detector, sPHENIX, based on a superconducting thin-coil solenoid surrounded by EM and Hadron calorimeters in order to measure jets as well as to improve on other hard-scattering measurements by taking advantage of high-rate calorimeter triggers and the much larger solid angle. Personally, I am fond of such a detector because the first thin coil superconducting solenoid detector at a collider was used in the CCOR experiment at CERN in 1977 (Fig. 19a) which made some of the major original

Figure 19. a) (left) CCOR thin coil superconducting solenoid, 70 cm radius, installed at the CERN ISR in 1977. b) (right) BABAR thin coil superconducting solenoid, 1.5 m radius, being shipped from Ansaldo in Italy to SLAC in 1997.

hard-scattering measurements [22, 53]. This year, sPHENIX got a big boost by the acquisition of the BABAR solenoid magnet from SLAC (Fig. 19b) which became available when the B-factory in Italy was cancelled. The conceptual design of the new experiment is well underway [54], with mid-rapidity, forward and eRHIC capability (Fig. 20). New collaborators are most welcome.

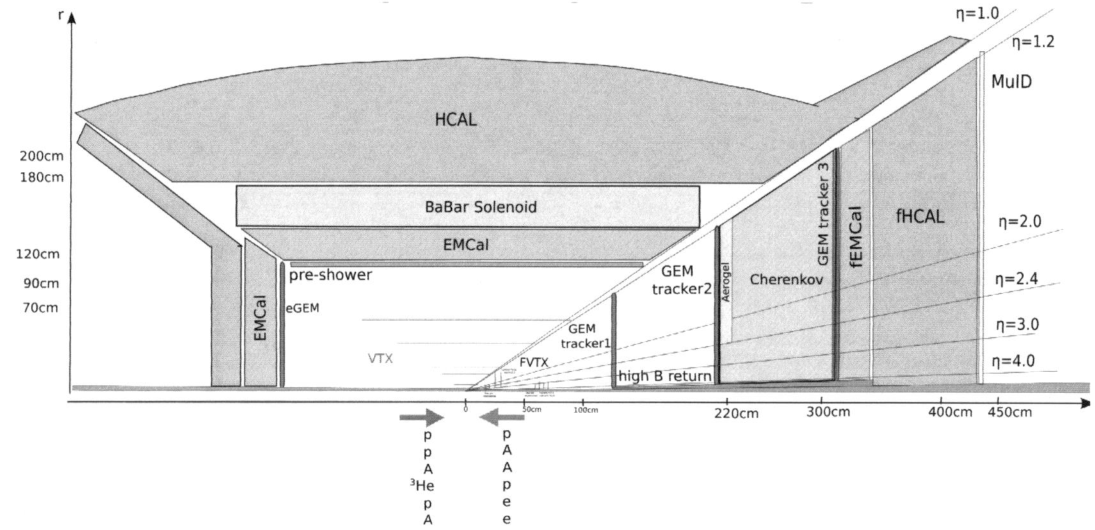

Figure 20. sPHENIX concept with forward detector [54].

References

[1] Orear J, Rosenfeld A H and Schluter R A 1950 *Nuclear Physics. A Course Given by Enrico Fermi at the University of Chicago* (The University of Chicago Press)
[2] Amsler C and others (Particle Data Group) 2008 *Phys. Lett.* B **667** 1
[3] Burke K M and others (JET Collaboration) 2013 Extracting jet transport coefficient from jet quenching at RHIC and LHC *Preprint* arXiv:1312.5003v2.
[4] Baier R, Schiff D and Zakharov B G 2000 *Annu. Rev. Nucl. Part. Sci.* **50** 37–69
[5] Baier R 2003 *Nucl. Phys.* A **715** 209c–218c
[6] Baier R, Dokshitzer Y L, Mueller A H, Peigné S and Schiff D 1997 *Nucl. Phys.* B **483** 291–320

[7] Adler S S and others (PHENIX Collaboration) 2006 *Phys. Rev.* D **74** 072002
[8] Adams J, Magestro D and others (STAR Collaboration) 2006 *Phys. Rev. Lett.* **97** 162301
[9] Adcox K and others (PHENIX Collaboration) 2002 *Phys. Rev. Lett.* **88** 022301
[10] Adler S S and others (PHENIX Collaboration) 2003 *Phys. Rev. Lett.* **91** 072303
[11] Adler S S and others (PHENIX Collaboration) 2007 *Phys. Rev.* C **76** 034904
[12] Adler S S and others (PHENIX Collaboration) 2003 *Phys. Rev. Lett.* **91** 172301
[13] Adler S S and others (PHENIX Collaboration) 2004 *Phys. Rev.* C **69** 034910
[14] Fritzsch H and Minkowski P 1977 *Phys. Lett.* B **69** 316–320
[15] Adler S S and others (PHENIX Collaboration) 2008 *Phys. Rev. Lett.* **101** 232301
[16] Adare A and others (PHENIX Collaboration) 2007 *Phys. Rev.* D **76** 051106(R)
[17] Adler S S and others (PHENIX Collaboration) 2005 *Phys. Rev. Lett.* **94** 232301
[18] Adare A and others (PHENIX Collaboration) 2012 *Phys. Rev.* **D 86** 072008
[19] Berman S M, Bjorken J D and Kogut J B 1971 *Phys. Rev.* D **4** 3388–3418
[20] Blankenbecler R, Brodsky S J and Gunion J F 1972 *Phys. Lett.* B **42** 461–465
[21] Cahalan R F, Geer K A, Kogut J and Susskind L 1975 *Phys. Rev.* D **11** 1199–1212
[22] Rak J and Tannenbaum M J 2013 *High p_T Physics in the Heavy Ion Era* (Cambridge University Press)
[23] Aamodt K and others (ALICE Collaboration) 2011 *Phys. Lett.* B **696** 30–39
[24] Adare A and others (PHENIX Collaboration) 2013 *Phys. Rev.* C **87** 034911
[25] Adare A and others (PHENIX Collaboration) 2012 *Phys. Rev. Lett.* **109** 152301
[26] Adare A and others (PHENIX Collaboration) 2010 *Phys. Rev. Lett.* **104** 252301
[27] Tannenbaum M J and others (PHENIX Collaboration) 2012 *Proc. 6th International Workshop on High-p_T Physics at the LHC* (*CERN-Proceedings* no 2012-001) ed Eskola K, van Leewen M, Levai P, Morsch A, Peitzmann T and Rak J (Geneva: CERN) pp 37–42
[28] Chatrchyan S and others (CMS Collaboration) 2011 *Phys. Rev.* C **84** 024906
[29] Chatrchyan S and others (CMS Collaboration) 2012 *Phys. Lett.* B **712** 176–197
[30] Adare A and others (PHENIX Collaboration) 2010 *Phys. Rev.* D **82** 072001
[31] Adare A and others (PHENIX Collaboration) 2013 *Phys. Rev. Lett.* **111** 032301
[32] Roland G and others (CMS Collaboration) 2013 *Nucl. Phys.* A **904-905** 43c–50c
[33] Adare A and others (PHENIX Collaboration) 2007 *Phys. Rev. Lett.* **98** 232302
[34] Mitchell J T and others (PHENIX Collaboration) 2007 *J. Phys.* G **34** S911–S914
[35] Alver B and Roland G 2010 *Phys. Rev.* C **81** 054905
[36] Adamcyzk L and others (STAR Collaboration) 2013 Jet-Hadron Correlations in $\sqrt{s_{NN}} = 200$ GeV Au+Au and p+p Collisions *Preprint* arXiv:1302.6184
[37] Caines H and others (STAR Collaboration) 2011 *J. Phys.* G **38** 124019
[38] Lacey R A, Taranenko A, Jia J, Ajitanand N N and Alexander J 2011 Scaling of the higher-order flow harmonics: implications for initial-eccentricity models and the “viscous horizon” *Preprint* arXiv:1105.3782v2
[39] Adare A and others (PHENIX Collaboration) 2011 *Phys. Rev. Lett* **107** 252301
[40] Eremin S and Voloshin S 2003 *Phys. Rev.* C **67** 064905
[41] Bialas A, Czyz W and Lesniak L 1982 *Phys. Rev. D* **25** 2328–2340
[42] Adler S S and others (PHENIX Collaboration) Transverse-energy distributions at midrapidity in p+p, d+Au, and Au+Au collisions at $\sqrt{s_{NN}} = 62.4-200$ GeV and implications for particle-production models *Preprint* arXiv:1312.6676, submitted to *Phys. Rev,* C
[43] Adare A and others (PHENIX Collaboration) 2013 *Phys. Rev.* C **88** 024906
[44] Adare A and others (PHENIX Collaboration) 2012 *Phys. Rev. Lett.* **109** 242301
[45] Cronin J W, Frisch H J, Shochet M J, Boymond J P, Piroué P A and Sumner R L 1975 *Phys. Rev.* D **11** 3105–3123
[46] Adler S S and others (PHENIX Collaboration) 2006 *Phys. Rev.* C **74** 024909
[47] Adler S S and others (PHENIX Collaboration) 2005 *Phys. Rev.* C **71** 051902(R)
[48] Adare A and others (PHENIX Collaboration) 2007 *Phys. Lett.* B **649** 359–369
[49] Fries R J, Müller B, Nonaka C and Bass S A 2003 *Phys. Rev. Lett.* **90** 202303
[50] Hwa R and Yang C B 2004 *Phys. Rev.* C **70** 037901
[51] Sahlmueller B and others (PHENIX Collaboration) 2013 *Nucl. Phys.* A **A904-905** 795c–798c
[52] Perepelitsa D V Inclusive jet production in ultrarelativistic proton-nucleus collisions Ph. D. Thesis, Columbia University, November, 2013
[53] Angelis A L S and others (CMOR Collaboration) 1988 *Nucl. Phys.* B **303** 569–606
[54] PHENIX Collaboration The sPHENIX MIE Proposal (submitted September 27,2013). Available at http://www.phenix.bnl.gov/plans

Journal of Physics: Conference Series **589** (2015) 012020 doi:10.1088/1742-6596/589/1/012020

Mid-rapidity D-meson production in pp, Pb–Pb and p–Pb collisions at the LHC

Jeremy Wilkinson for the ALICE Collaboration
Physikalisches Institut, Ruprecht-Karls-Universität Heidelberg, Heidelberg, Germany
E-mail: jwilkinson@physi.uni-heidelberg.de

Abstract. We present the recent results for D-meson production measured by the ALICE collaboration in pp collisions at $\sqrt{s} = 7$ TeV, Pb–Pb collisions at $\sqrt{s_{\rm NN}} = 2.76$ TeV and p–Pb collisions at $\sqrt{s_{\rm NN}} = 5.02$ TeV.

1. Introduction

Measurements of heavy-flavour production at a variety of energies and in various colliding systems offer a unique probe into the properties of the quark-gluon plasma (QGP) formed in heavy-ion collisions. Due to their large mass, charm and beauty quarks are produced in the early stages of the collision through hard partonic interactions. This means that they maintain their identity throughout the full evolution of the collision, thereby serving as a tagged probe.

The ALICE experiment studies heavy-flavour production in part through the reconstruction of D mesons produced in the collisions. This is performed in three collision systems, each giving different insights into the overall mechanisms of charm production. Proton-proton collisions serve as a fundamental test of perturbative quantum chromodynamic (pQCD) calculations, as well as providing an essential reference spectrum with which heavy-ion results can be compared; Pb–Pb collisions give the possibility of studying the properties of strongly-interacting matter at high temperatures and densities; and p–Pb collisions allow us to disentangle cold nuclear effects (such as $k_{\rm T}$ broadening, the nuclear modification of parton distribution functions, and parton saturation at low x) from hot nuclear matter effects arising in the medium.

The medium's properties can be studied using two of the experimental observables extracted from heavy-ion results, namely the nuclear modification factor $R_{\rm AA}$ and the elliptic flow parameter v_2. The $R_{\rm AA}$ measures modifications of the transverse momentum distributions of D mesons, and is sensitive to the energy loss of charm quarks due to e.g. gluon bremsstrahlung and elastic collisions with other constituents of the medium. It is defined as

$$R_{\rm AA}(p_{\rm T}) = \frac{1}{\langle N_{\rm coll}\rangle} \frac{{\rm d}N_{\rm AA}/{\rm d}p_{\rm T}}{{\rm d}N_{\rm pp}/{\rm d}p_{\rm T}}, \qquad (1)$$

where $\langle N_{\rm coll}\rangle$ is the number of binary collisions, $N_{\rm AA}$ is the measured yield in Pb–Pb collisions and $N_{\rm pp}$ is the yield in pp collisions at the same collision energy. It is expected that $R_{\rm AA}$ should be equal to unity if there are no nuclear effects present. In addition, the energy loss depends on colour charge and the quark-mass-dependent dead cone effect [1]. Due to this, there is an expected hierarchy of suppression between different parton species. This cannot be measured

Journal of Physics: Conference Series **589** (2015) 012020 doi:10.1088/1742-6596/589/1/012020

directly, but will also alter the production of their decay products. It is expected, therefore, that $R_{\mathrm{AA}}(\pi) < R_{\mathrm{AA}}(\mathrm{D}) < R_{\mathrm{AA}}(\mathrm{B})$.

The elliptic flow v_2 is a measure of the momentum anisotropy of the particles produced in an interaction. Such anisotropy occurs especially in semi-peripheral heavy-ion collisions due to the overlap region of the nuclei being spatially anisotropic. This can be expressed with a Fourier expansion of the distribution of produced particles as a function of the azimuthal angle φ relative to the reaction plane Ψ_{RP}:

$$\frac{\mathrm{d}N}{\mathrm{d}(\varphi - \Psi_{\mathrm{RP}})} = \frac{N_0}{2\pi}(1 + 2v_1\cos(\varphi - \Psi_{\mathrm{RP}}) + 2v_2\cos(2(\varphi - \Psi_{\mathrm{RP}})) + ...), \qquad (2)$$

with v_n representing the different harmonics of the expansion. The reaction plane is defined so as to contain both the impact parameter and the beam axis. At low and intermediate p_{T}, the value of v_2 offers an insight into the level of thermalisation of charm quarks in the medium, while at higher p_{T} it allows us to study the path-length dependence of heavy-quark energy loss.

In these proceedings, we will discuss the ALICE measurements of D-meson production in pp collisions at $\sqrt{s} = 7$ TeV, Pb–Pb collisions at $\sqrt{s_{\mathrm{NN}}} = 2.76$ TeV, and p–Pb collisions at $\sqrt{s_{\mathrm{NN}}} = 5.02$ TeV.

2. Analysis strategy

The analysis of D mesons reconstructed inside the ALICE central barrel makes use of the detector's excellent tracking, vertexing and particle identification (PID) capabilities.

For this analysis, we fully reconstruct D mesons (and their respective charge conjugates) in their hadronic decay channels, with branching ratios $\mathcal{B}$ and mean proper decay lengths $c\tau$ as follows [2]:

$$\mathrm{D}^0 \to \mathrm{K}^-\pi^+(\mathcal{B} = 3.88 \pm 0.05\%, c\tau = 123\mu\mathrm{m});$$
$$\mathrm{D}^+ \to \mathrm{K}^-\pi^+\pi^+(\mathcal{B} = 9.13 \pm 0.19\%, c\tau \approx 312\mu\mathrm{m});$$
$$\mathrm{D}^{*+} \to \mathrm{D}^0\pi^+(\mathcal{B} = 67.7 \pm 0.5\%) \to \mathrm{K}^-\pi^+\pi^+;$$
$$\mathrm{D}_\mathrm{s}^+ \to \varphi\pi^+ \to \mathrm{K}^+\mathrm{K}^-\pi^+(\mathcal{B} = 2.28 \pm 0.12\%, c\tau \approx 150\mu\mathrm{m}).$$

The analysis of D-meson production mainly makes use of four detector systems: the Inner Tracking System (ITS), the VZERO detector, the Time Projection Chamber (TPC) and the Time-of-Flight detector (TOF). The ITS, formed of a series of silicon pixel, strip and drift detectors, offers high-resolution tracking and vertexing at mid-rapidity ($|\eta| < 0.9$). The VZERO detector is a scintillator detector covering the pseudorapidity ranges $2.8 < \eta < 5.1$ (VZEROA) and $-3.7 < \eta < -1.7$ (VZEROC), and is used for triggering and centrality determination. The TPC is a gas detector, which surrounds the ITS in the region $|\eta| < 0.9$ and offers further tracking as well as PID via measurements of specific energy loss. Finally, the TOF detector offers PID at mid-rapidity via measurements of particles' times of flight.

As D mesons have a non-zero lifetime, they typically travel some distance from the interaction point before decaying. This allows us to make topological selections of possible decay daughters based on e.g. the impact parameter (the distance of closest approach between the daughter track and the primary vertex) and the pointing angle (the angle between the meson flight line and its reconstructed momentum). In the case of the D^{*+} meson, the lifetime is eight orders of magnitude shorter than the other species, making a direct topological selection impossible. Instead we pair a D^0 candidate with a $\pi^\pm$ at the primary vertex to reconstruct these.

In tandem with the topological selections, we also use the PID capabilities of the TPC and TOF to identify the daughter tracks based on the measured responses in these detectors. Accepted candidates are then used to fill p_{T}-binned invariant mass spectra, from which the raw yields are extracted using a fit function composed of a Gaussian function for the signal and an

Journal of Physics: Conference Series **589** (2015) 012020 doi:10.1088/1742-6596/589/1/012020

exponential function for the background. The mean of the signal peak is typically found to be compatible with to the PDG mass value [2]. The width of the peak, which for D^0 and D^+ varies between ~6 and ~20 MeV/c^2, is well reproduced by Monte Carlo (MC) simulations. These yields are then corrected for their detection efficiencies, which are determined using MC simulations of the specific collision system.

Finally, it is necessary to subtract feed-down due to the decays of B mesons produced in the collisions. This subtraction is performed using FONLL [3] calculations, the reconstruction and selection efficiency of D mesons from B feed-down, and a hypothesis of the R_{AA} (R_{pPb}) of D mesons arising from B-meson decays in Pb–Pb (p–Pb) collisions.

3. Results in pp collisions

The analysis in pp collisions at $\sqrt{s} = 7$ TeV was performed using a data sample taken in 2010 with a minimum-bias trigger. The analysed data corresponded to an integrated luminosity of 5 nb^{-1}. This measurement was made at mid-rapidity, in the region $1 < p_T < 24$ GeV/c. The differential cross sections were reported in [4], and the result for D^{*+} production is shown in figure 1. The measured cross sections are described within uncertainties by FONLL [3] and GM-VFNS [5] pQCD calculations, with the experimental results matching the upper end of the uncertainty bands in the FONLL calculations.

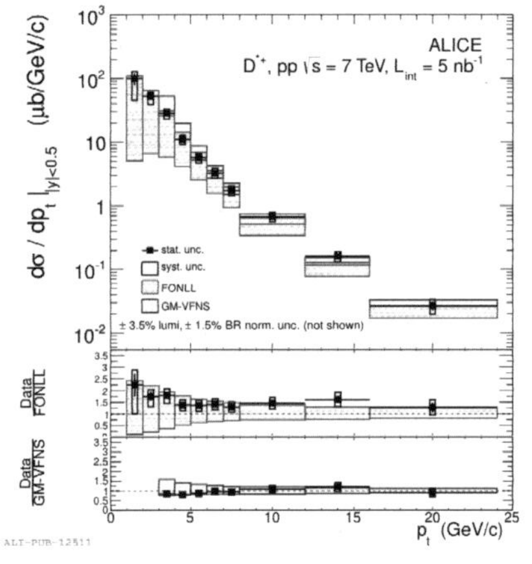

Figure 1. p_T-differential cross section of D^{*+} production in pp collisions at $\sqrt{s} = 7$ TeV, compared with results from FONLL [3] and GM-VFNS [5] pQCD calculations [4].

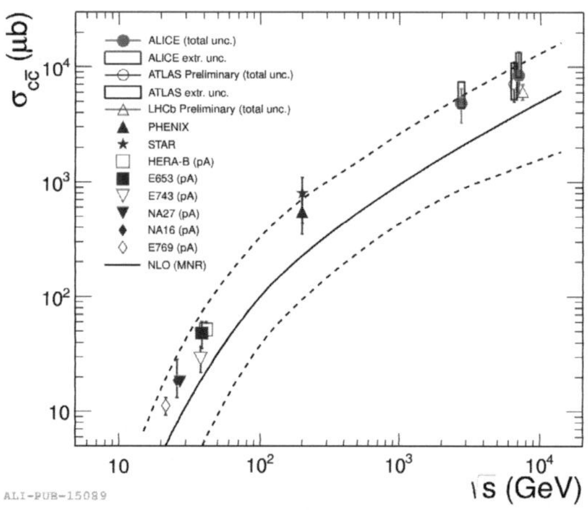

Figure 2. Total charm cross section measured in pp collisions with ALICE as a function of $\sqrt{s}$, in comparison with other experiments and MNR [6] calculations [7].

Further measurements were also taken for pp collisions at $\sqrt{s} = 2.76$ TeV, which was the centre-of-mass energy used for Pb–Pb collisions. These results were reported in [7]. Since the statistics collected at this energy were fewer than those recorded at $\sqrt{s} = 7$ TeV, these were not used as the pp reference for the Pb–Pb results, instead being used mainly to verify the validity of an energy scaling of the 7 TeV results to $\sqrt{s} = 2.76$ TeV using FONLL [8].

In addition, the results for 7 TeV and 2.76 TeV pp collisions were extrapolated to full phase space using FONLL calculations, in order to extract the total $c\bar{c}$ cross section in pp at these energies. The results of these extrapolations are shown in figure 2, alongside results from other experiments and next-to-leading-order pQCD calculations using the MNR framework [6]. The result at 7 TeV is in agreement with those from the ATLAS and LHCb collaborations, and both of the ALICE points follow the trends exhibited by both the NLO predictions and other experimental results spanning three orders of magnitude in energy. One thing to note here is that all of the results lie within the upper part of the uncertainty band of the predictions; this is possibly due to the central calculation overestimating the bare charm quark mass ($m_c = 1.5$ GeV/c^2).

4. Results in Pb–Pb collisions

The Pb–Pb data at $\sqrt{s_{\mathrm{NN}}} = 2.76$ TeV were taken in 2010 and 2011, using both minimum-bias and centrality-triggered data. In 2010, a data sample of 2 μb^{-1} was taken with a minimum-bias trigger, while in 2011 an online selection based on VZERO was used to enhance the sample of central (28 μb^{-1}) and mid-central (6 μb^{-1}) collisions. In order to provide the reference spectrum at $\sqrt{s} = 2.76$ TeV, the high-statistics 7 TeV data were scaled to 2.76 TeV using FONLL, as discussed in section 3. The nuclear modification factors of D^0, D^+ and D^{*+} mesons using the 2010 data sample were published in [9], and the v_2 results for these from the 2011 data sample were published in [10].

The R_{AA} for the 7.5% most-central Pb–Pb collisions was measured for all four D-meson species in different p_{T} regions spanning $1 < p_{\mathrm{T}} < 36$ GeV/c. The measurements for each species are shown in figure 3. In figure 4, the average R_{AA} of D^0, D^+ and D^{*+} mesons in the centrality range 0-20%, taken from [9], is compared with different model calculations implementing in-medium parton energy loss.

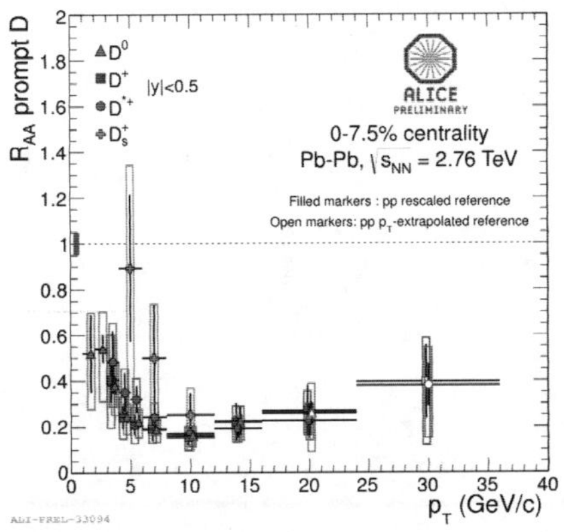

Figure 3. R_{AA} of D^0, D^+, D^{*+}, and D_s^+ mesons in central Pb–Pb collisions at $\sqrt{s_{\mathrm{NN}}} = 2.76$ TeV.

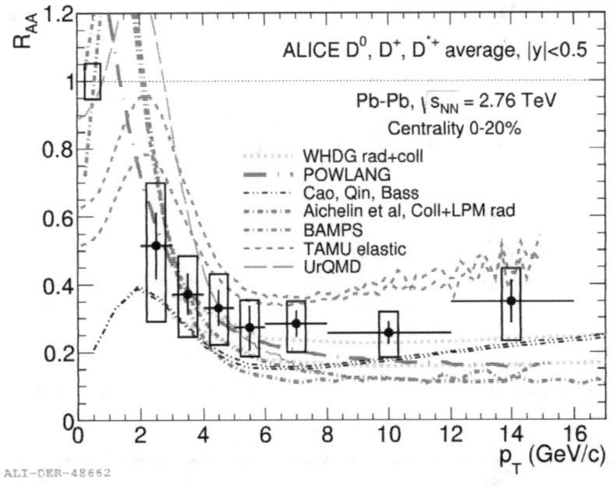

Figure 4. Average R_{AA} for D^0, D^+, and D^{*+} mesons, compared with model predictions [11–19].

The results for D^0, D^+ and D^{*+} agree with one another within uncertainties, and display a particularly high level of suppression at $p_{\mathrm{T}} > 6$ GeV/c. The R_{AA} of D_s^+, measured in the range $4 < p_{\mathrm{T}} < 12$ GeV/c, is also displayed. At present, the uncertainties are too large to conclude whether or not there is a lower level of suppression for D_s^+ at low to intermediate p_{T} in comparison with the other D mesons. Such lifting of the strangeness suppression in Pb–Pb compared to pp is theorised to possibly happen due to charm recombination in the medium [20, 21].

The centrality-dependent R_{AA} was also determined for D^0, D^+, and D^{*+}. The preliminary results for D^0 at low p_{T} are shown in figure 5, and at higher p_{T} for all three species in figure 6. From these it can be seen that the centrality-dependent behaviour of the R_{AA} varies largely with p_{T} at low momenta, eventually stabilising at higher p_{T}. In addition, there is no significant difference in the centrality dependence of the suppression between the three D-meson species.

Journal of Physics: Conference Series **589** (2015) 012020 doi:10.1088/1742-6596/589/1/012020

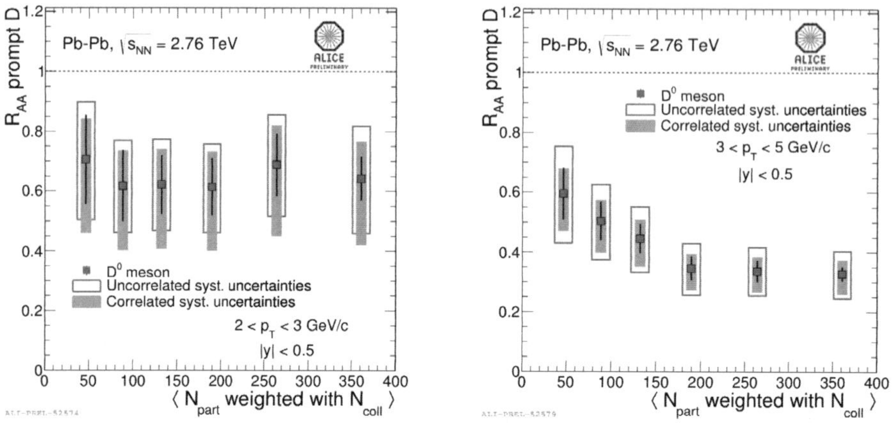

Figure 5. R_{AA} as a function of centrality for D^0 mesons in Pb–Pb collisions at $\sqrt{s_{NN}}$ = 2.76 TeV, for $2 < p_T < 3$ GeV/c (left) and $3 < p_T < 5$ GeV/c (right).

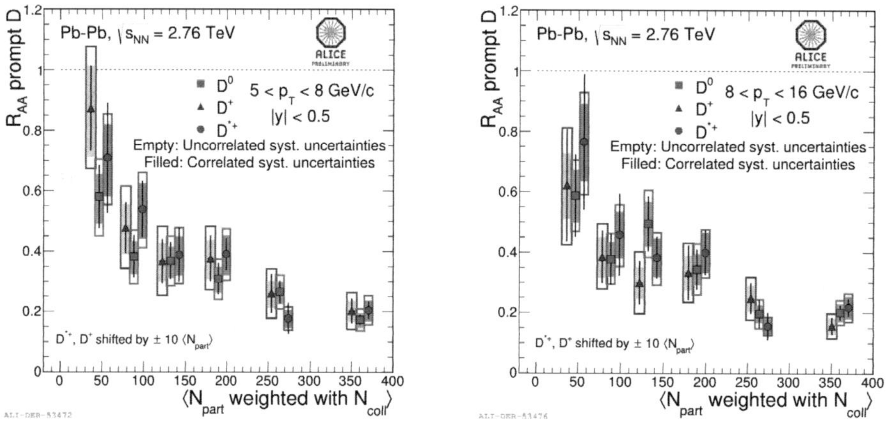

Figure 6. R_{AA} as a function of centrality for D^0, D^+ and D^{*+} mesons in Pb–Pb collisions at $\sqrt{s_{NN}} = 2.76$ TeV, for $5 < p_T < 8$ GeV/c (left) and $8 < p_T < 16$ GeV/c (right).

This result was also compared with that for non-prompt J/ψ mesons (i.e. J/ψ particles from the decay of B mesons) measured by the CMS collaboration [22]. It should be noted that the CMS measurements were taken in different rapidity regions to the ALICE results. For direct comparisons to be made, the D-meson p_T intervals were chosen such that they represented similar kinematic regions, using kinematic simulations of $B \to J/\psi + X$. These comparisons are shown in figure 7. The results indicate that there is a greater level of suppression for charm quarks than for beauty, as expected from the predicted mass dependence discussed in section 1.

The v_2 of D mesons at mid-rapidity was measured in the centrality range 30-50% [10] using the event-plane method, defined as follows:

$$v_2 = \frac{1}{R_2} \frac{\pi}{4} \frac{N_{\text{in-plane}} - N_{\text{out-of-plane}}}{N_{\text{in-plane}} + N_{\text{out-of-plane}}}, \tag{3}$$

where the factor $\frac{1}{R_2}$ is used to correct for the finite resolution involved in determining the reaction plane. $N_{\text{in-plane}}$ and $N_{\text{out-of-plane}}$ refer to the raw yields in two azimuthal regions relative to the reaction plane: in-plane ($-\frac{\pi}{4} < \Delta\varphi < \frac{\pi}{4}$ and $\frac{3\pi}{4} < \Delta\varphi < \frac{5\pi}{4}$) and out-of-plane ($\frac{\pi}{4} < \Delta\varphi < \frac{3\pi}{4}$

Journal of Physics: Conference Series **589** (2015) 012020 doi:10.1088/1742-6596/589/1/012020

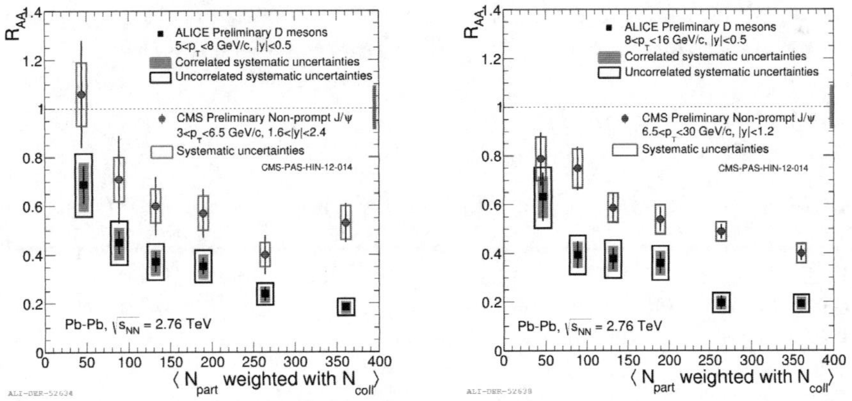

Figure 7. Average R_{AA} as a function of centrality for D^0, D^+ and D^{*+} mesons in Pb–Pb collisions at $\sqrt{s_{\mathrm{NN}}} = 2.76$ TeV, compared with the R_{AA} of non-prompt J/ψ mesons measured by the CMS collaboration in similar kinematic regions.

and $\frac{5\pi}{4} < \Delta\varphi < \frac{7\pi}{4}$).

The average v_2 for D^0, D^+ and D^{*+} mesons is shown in figure 8, compared with the measured v_2 of charged particles. The D-meson v_2 is significantly greater than zero at low p_{T}, with a greater than 5σ significance at $2 < p_{\mathrm{T}} < 6$ GeV/c, as well as being compatible with the overall v_2 of charged particles (which is dominated by light-flavour hadrons). This suggests that low-momentum charm quarks take part in the collective motion of the system.

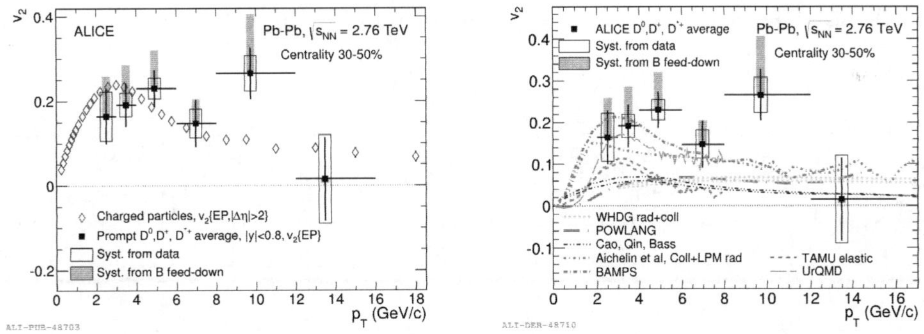

Figure 8. Average v_2 of D^0, D^+ and D^{*+} mesons as a function of p_{T} for Pb–Pb collisions at $\sqrt{s_{\mathrm{NN}}} = 2.76$ TeV in the 30-50% centrality range, compared with the v_2 of charged particles (left) and with model predictions (right) [11–19].

The combination of the R_{AA} and v_2 measurements provides a strict constraint for the predictions given by models. By contrasting the comparison of R_{AA} to models in figure 4 with that of v_2 with model predictions in figure 8, it can be seen that some of the currently-available models are able to reproduce either the R_{AA} or the v_2 measured in experiment, however a simultaneous description of both of these results remains a challenge.

5. Results in p–Pb collisions

The data for p–Pb collisions at $\sqrt{s_{\mathrm{NN}}} = 5.02$ TeV used in the D-meson analysis were taken during a dedicated run at the start of 2013, just prior to start of the first Long Shutdown (LS1) of the LHC. The data were taken with a minimum-bias trigger, with an integrated luminosity of

Journal of Physics: Conference Series **589** (2015) 012020 doi:10.1088/1742-6596/589/1/012020

49 μb^{-1}. Once the differential cross sections were determined, they were used to compute the modification factor R_{pPb} using a similar method to the R_{AA} calculation:

$$R_{\mathrm{pPb}}(p_{\mathrm{T}}) = \frac{\mathrm{d}\sigma_{\mathrm{pPb}}/\mathrm{d}p_{\mathrm{T}}}{A \times (\mathrm{d}\sigma_{\mathrm{pp}}/\mathrm{d}p_{\mathrm{T}})}, \tag{4}$$

where σ_{pPb} is the cross section in p–Pb collisions, σ_{pp} is the cross section in pp collisions at the same centre-of-mass energy, and A is the atomic mass number of the Pb nucleus. The preliminary R_{pPb} measurements for D^0, D^+, D^{*+}, and D_s^+ mesons are shown in figure 9. The measurements for all D-meson species, including D_s^+, are compatible with one another, as well as being consistent with unity within uncertainties.

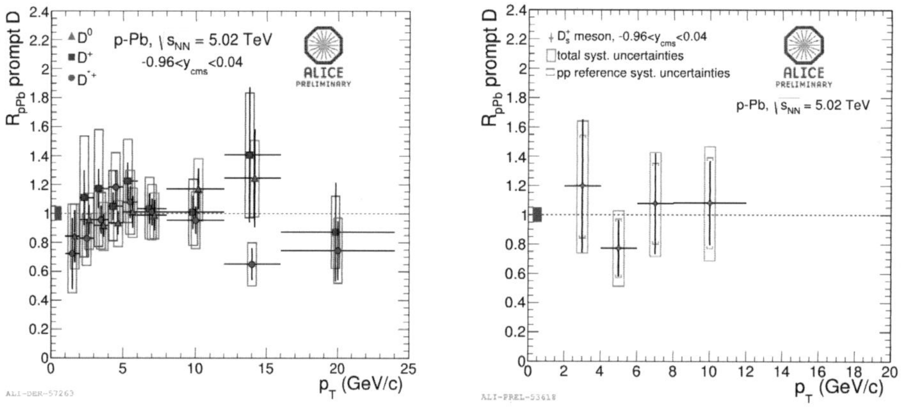

Figure 9. R_{pPb} of D^0, D^+, D^{*+} (left) and D_s^+ (right) mesons as a function of p_{T} in p–Pb collisions at $\sqrt{s_{\mathrm{NN}}} = 5.02$ TeV.

The R_{pPb} measurements for D^0, D^+ and D^{*+} mesons were averaged and compared with results from two models, namely next-to-leading order pQCD calculations using the MNR framework [6] with EPS09 for the description of the nuclear PDFs [23]; and the Colour Glass Condensate (CGC) framework [24]. This comparison is shown in figure 10. It can be seen that, within uncertainties, the experimental results do not favour one prediction over the other.

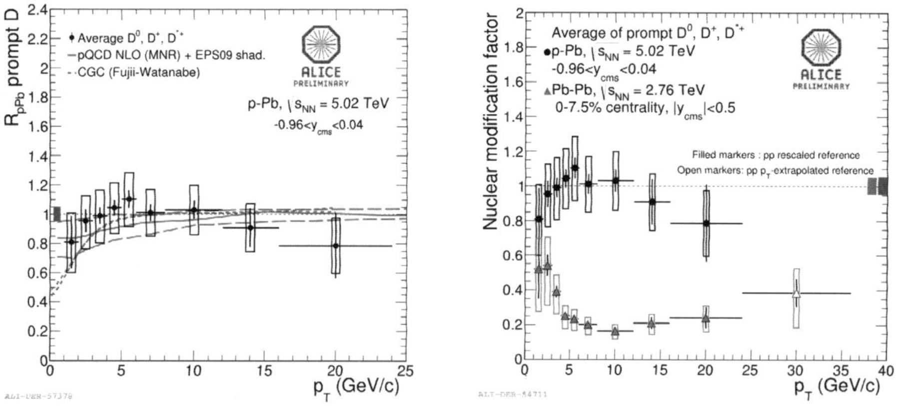

Figure 10. Average R_{pPb} of D^0, D^+ and D^{*+} mesons as a function of p_{T} at $\sqrt{s_{\mathrm{NN}}} = 5.02$ TeV, compared with models [6, 23] (left) and R_{AA} (right)

Journal of Physics: Conference Series **589** (2015) 012020 doi:10.1088/1742-6596/589/1/012020

Finally, in figure 10, we also compare the average R_{pPb} with the average R_{AA} measured for D mesons, which was discussed in section 4. We find that, although the centre-of-mass energies of the two systems differ, there is a clear suppression in Pb–Pb collisions compared with p–Pb for $p_{\text{T}} > 4$ GeV/c. This indicates that final-state effects such as partonic energy loss in the medium have a far more dominant role in the suppression of D-meson production at higher p_{T} than initial-state effects such as nuclear shadowing.

6. Conclusions and Outlook

The reconstructed decays of D mesons were measured at mid-rapidity in pp, Pb–Pb and p–Pb collisions. The pp measurements, which were used as reference spectra for the heavy-ion results, were found to be in agreement with pQCD calculations, and extrapolations of the cross sections to full phase space were consistent with the trends exhibited by both experiment and theory across three orders of magnitude in energy.

In Pb–Pb collisions, it was determined that the level of suppression experienced by non-strange D mesons is similar between species, with a significant suppression exhibited at high p_{T}. This suppression can be understood as an indication of in-medium parton energy loss. Further study will be required before conclusions can be drawn as to whether there is an enhancement of the D_s^+ yield relative to non-strange D mesons at low p_{T}. In addition, as expected due to the dead-cone effect, the R_{AA} of D mesons was found to be lower than that of B mesons in the range $5 < p_{\text{T}} < 16$ GeV/c. Finally, a significantly non-zero v_2 for D mesons was measured at low p_{T}, suggesting that charm quarks take part in the collective motion of the medium. The results from p–Pb collisions show the R_{pPb} to be consistent with unity, as well as being compatible with models of initial-state effects. This suggests that the suppression visible at high p_{T} in Pb–Pb collisions primarily occurs due to hot nuclear matter effects rather than initial-state effects.

The rapidity and multiplicity dependence of D-meson production in p–Pb collisions are currently being studied. In 2015, once the current Long Shutdown of the LHC is complete, ALICE will resume taking data at a higher $\sqrt{s_{\text{NN}}}$, allowing further refinement of the results with an expected factor of 10 increase of statistics prior to the start of the second Long Shutdown.

References

[1] Dokshitzer Y L and Kharzeev D E 2001 *Phys. Lett. B* **519** 199–206
[2] Beringer J *et al.* (Particle Data Group) 2012 *Phys. Rev. D* **86** 010001
[3] Cacciari M, Greco M and Nason P 1998 *JHEP* **9805** 007
[4] Abelev B *et al.* (ALICE Collaboration) 2012 *JHEP* **1201** 128
[5] Kniehl B A *et al.* 2012 *Preprint* arXiv:1202.0439 [hep–ph]
[6] Mangano M, Nason P and Ridolfi G 1992 *Nucl. Phys. B* **373** 295–345
[7] Abelev B *et al.* (ALICE Collaboration) 2012 *JHEP* **1207** 191
[8] Averbeck R *et al.* 2011 *Preprint* arXiv:1107.3243 [hep–ph]
[9] Abelev B *et al.* (ALICE Collaboration) 2012 *JHEP* **09** 112
[10] Abelev B *et al.* (ALICE Collaboration) 2013 *Phys. Rev. Lett.* **111** 102301
[11] Horowitz W A and Gyulassy M 2011 *J. Phys. G* **38** 124114
[12] Alberico W M *et al.* 2011 *Eur. Phys. J. C* **71** 1666
[13] Cao S, Qin G Y and Bass S A 2013 *Phys. Rev. C* **88** 044907
[14] Gossiaux P B, Bierkandt R and Aichelin J 2009 *Phys. Rev. C* **79** 044906
[15] Gossiaux P B *et al.* 2010 *J. Phys. G* **37** 094019
[16] Fochler O *et al.* 2011 *J. Phys. G* **38** 124152
[17] He M, Fries R J and Rapp R 2012 *Phys. Rev. C* **86** 014903
[18] Lang T *et al.* 2012 *Preprint* arXiv:1211.6912 [hep–ph]
[19] Lang T *et al.* 2012 *Preprint* arXiv:1212.0696 [hep–ph]
[20] He M, Fries R J and Rapp R 2013 *Phys. Rev. Lett.* **110** 112301
[21] Kuznetsova I and Rafelski J 2007 *Eur. Phys. J. C* **51** 113–133
[22] CMS Collaboration 2012 CMS–PAS–HIN–12–014
[23] Eskola K J, Paukkunen H and Salgado C A 2009 *JHEP* **0904** 065
[24] Fujii H and Watanabe K 2013 *Preprint* arXiv:1308.1258 [hep–ph]

Journal of Physics: Conference Series **589** (2015) 012021 doi:10.1088/1742-6596/589/1/012021

Analysis of proton-lead data via re-weighting

P Zurita

Departamento de Física de Partículas and IGFAE, Universidade de Santiago de Compostela, E-15706 Santiago de Compostela, Galicia, Spain.

E-mail: pia.zurita@usc.es

Abstract.
The recent proton-lead run at the LHC shall provide new information on the partonic behaviour within the nuclear medium. At LHC energies the dominant contribution comes from gluon-initiated processes that is, the least well constrained parton density. Therefore it is important to profit from any information that new data can provide us. A time-saving alternative to performing a global fit is the use of Bayesian inference, a powerful tool to realize the impact of data into a set of PDFs independently of the original fitters. In this work we apply the Bayesian re-weighting technique to analyze pseudo data for LHC kinematics in Drell-Yan and hadro-production processes. A set of Monte Carlo replicas for EPS09 is released in a public code for general use.

1. Introduction

The precise determination of nuclear PDFs (nPDFs) has become an important asset in the light of the heavy-ion collisions at the LHC. However, due to the scarce amount of data from previous experiments and their limited kinematical coverage, the parton densities inside a nucleus are far from being well known. In particular, the nuclear gluon distribution has no real constrains. In this respect, the measurements from the LHC proton-lead run have the potential [1, 2, 3, 4, 5, 6, 7] to pinpoint the gluon behaviour in heavy nuclei.

The usual way to obtain information on the PDFs from a new experiment is to include the new results in a global fit, together with all previously analyzed data. Several choices must be done and variations of the fitting strategy have to be explored, turning it into a time-consuming process. One alternative to avoid performing a new fit was developed for proton PDFs based on the Monte Carlo method [8, 9]. By means of Bayesian inference one can quantitatively study the impact of new data on the central values and uncertainties of the PDFs.

In this contribution we summarize the results presented in [10]. There we apply the Bayesian re-weighting technique to explore the constraining potential of the LHC proton-lead run data on EPS09 [11] nuclear PDFs, taking two processes as example: Drell-Yan production and charged hadron inclusive production. The analyses are performed using pseudo-data based on the collinear DGLAP framework and also in the Color Glass Condensate (CGC) framework [12] for the case of hadro-production. The set of Monte Carlo replicas of EPS09 generated for this study is released as a public computer code.

In section 2 we present the main features of the re-weighting procedure and construct a Monte Carlo version of the EPS09 set. The impact of the new data on EPS09 nPDFS for Drell-Yan production and charged hadron production is studied in section 3. In the latter case, we also

Journal of Physics: Conference Series **589** (2015) 012021 doi:10.1088/1742-6596/589/1/012021

explore the discriminating power between DGLAP and CGC scenarios. Section 4 summarizes our results.

2. Bayesian re-weighting

Nowadays two methods are used in order to obtain the PDF uncertainties: the Hessian approach (used also for the nuclear PDFs) and the Monte Carlo approach (by the NNPDF Collaboration [13, 14]). In the latter the underlying PDF probability density $\mathcal{P}$ is sampled by generating, through a Monte Carlo procedure, an ensemble of N_{rep} PDFs replicas f_k, $k = 1, ..., N_{rep}$, each fitted to a replica of the experimental data. Then any quantity $\mathcal{O}[f]$ depending on the PDFs can be evaluated by

$$\langle \mathcal{O} \rangle = \frac{1}{N_{\rm rep}} \sum_{k=1}^{N_{\rm rep}} \mathcal{O}[f_k]\,. \tag{1}$$

If we now want to include the information of a new measurement consisting of n points with covariance matrix cov_{ij}, $y = y_1, y_2, ..., y_n$, we can use Bayesian inference [8, 9] to update the original probability distribution $\mathcal{P}_{old}(f)$ to a new probability distribution $\mathcal{P}_{new}(f)$. This is achieved by computing a weight w_k for each individual replica f_k, which quantifies its agreement with the new data. In terms of the χ^2_k between the original theory predictions for the $k-th$ replica and the new experimental measurement, the weights are

$$w_k = \frac{(\chi^2_k)^{\frac{1}{2}(n-1)} e^{-\chi^2_k/2}}{\frac{1}{N_{\rm rep}} \sum_{k=1}^{N_{\rm rep}} (\chi^2_k)^{\frac{1}{2}(n-1)} e^{-\chi^2_k/2}}\,, \tag{2}$$

So that now the analog of (eq.1) is

$$\langle \mathcal{O} \rangle_{\rm new} = \frac{1}{N_{\rm rep}} \sum_{k=1}^{N_{\rm rep}} w_k \mathcal{O}[f_k]\,. \tag{3}$$

As long as the new data is not too constraining, the Bayesian re-weighting is equivalent to a full new fit, though the former loses some statistical efficiency with respect to the latter. The replicas with very small weights will become irrelevant when averaging and the accuracy of the representation of the underlying $\mathcal{P}_{new}(f)$ will decrease. We quantify the efficiency loss by computing N_{eff}, the effective number of replicas after re-weighting:

$$N_{\rm eff} \equiv \exp\left\{ \frac{1}{N_{\rm rep}} \sum_{k=1}^{N_{\rm rep}} w_k \log(N_{\rm rep}/w_k) \right\}\,. \tag{4}$$

The accuracy of the re-weighted fit is the same that would be obtained if a new fit with N_{eff} replicas were to be performed. When $N_{eff} \ll N_{rep}$ the re-weighting method becomes unreliable and a full refit is mandatory.

The original re-weighting was developed for PDF sets based on the Monte Carlo approach. However when using a nuclear PDFs set (Hessian uncertainties), we must first generate the Monte Carlo replicas [15]. For this we took the central value of each EPS09 [11] parton (f_0) and shifted it according to

$$f_k(x, Q^2) = f_0(x, Q^2) + \sum_{i}^{N_{\rm eig}} \{f_i^{\pm}(x, Q^2) - f_0(x, Q^2)\} |r_{k,i}|\,, \quad (k = 1, \dots, N_{\rm rep})\,, \tag{5}$$

with N_{eig} the number of pairs of Hessian eigenvectors, $r_{k,i}$ random numbers from a Gaussian distribution of mean zero and variance one, and $f_{\pm i}$ the nPDF corresponding to the eigenvector $S_{\pm i}$. A symmetric version can be obtained by averaging over each pair of eigenvectors.

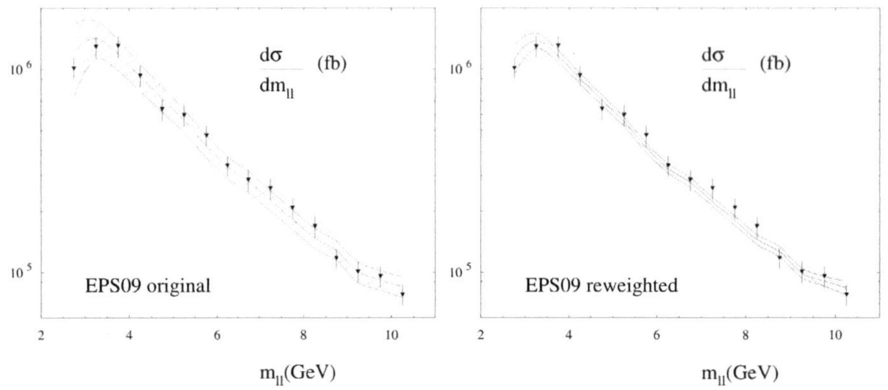

Figure 1. The predictions for the Drell-Yan differential cross sections with EPS09, compared to pseudo-data as a function of the invariant lepton pair mass m_{ll} at central rapidity before (left) and after (right) PDF re-weighting.

3. Results

3.1. Constraints from Drell-Yan production

first we analyze the impact on nuclear PDFs of Drell-Yan production in proton-lead collisions at the LHC, which should provide crucial information to further constrain the gluon density at small and medium x as well as the sea quark distributions. We are particularly interested in low-mass Drell-Yan production due to the enhanced sensitivity to nuclear effects [16].

In our exercise we consider only Drell-Yan cross sections for $|\eta| < 4$, without kinematical cuts on $p_{T,l}$. We generate both the pseudo-data and the predictions for DY production based on the $N_{rep} = 1000$ replicas of the MC version of EPS09 for $m_{ll} < 12GeV$ by use of the MCFM code [17], modified to take into account the fact that one of the initial particles is a lead nucleus. The pseudo-data was computed from the central values of EPS09 and adding the corresponding statistical fluctuations and a total uncorrelated 8% systematic uncertainty. In what follows we only consider the PDF uncertainties coming from EPS09.

In figure 1 we show the pseudo-data and the predictions before (left) and after (right) re-weighting. Once the new data is included the uncertainty band of the theory prediction is reduced, without affecting the central value, as expected for consistent data. In table 1 we present the effective number of replicas N_{eff}, and the values for the mean χ^2 and the average over replicas $\langle\chi^2\rangle$ per data point before and after the re-weighting. We see that the χ^2/n is $\mathcal{O}(1)$, as expected by the use of consistent pseudo-data. About half of the replicas are disfavored by the pseudo-data: these are suppressed by re-weighting, reducing the value of $\langle\chi^2\rangle$.

Table 1. χ^2/n and $\langle\chi^2\rangle/n$ values before and after the reweighting of EPS with the Drell-Yan pPb pseudo-data, with $n = 16$ points. The effective number of replicas $N_{\rm eff}$ is also provided for the re-weighting case.

	χ^2/n	$\langle\chi^2\rangle/n$	$N_{\rm eff}$
Original	0.64	2.68	-
Reweighted	0.59	0.96	539

By applying the re-weighting to the parton densities, we can assess the impact of the DY pseudo-data on the nuclear PDFs. While for the valence distributions there is no significant change, the sea distributions present a reduction on uncertainty bands in the $x < 10^{-2}$ region (plots not shown). The one parton that varies distinctly is the gluon, as can be seen in figure 2: the pseudo-data seems to favor more suppressed gluons in the $x < 10^{-2}$ region, and the estimated reduction of the uncertainty is up to 50%.

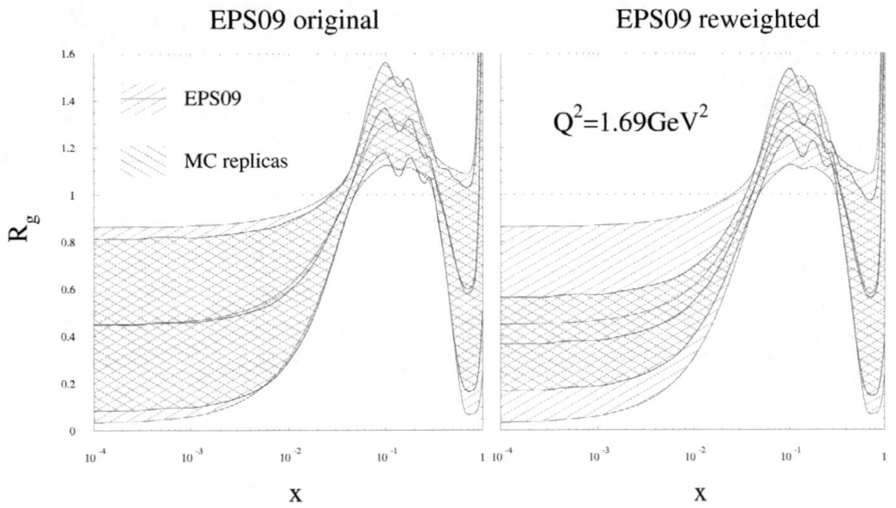

Figure 2. EPS09 nuclear ratio $R_g^A(x, Q^2)$ for $Q^2 = 1.69$ GeV^2, both in the Hessian and in the Monte Carlo versions, before (left) and after (right) the reweighting with LHC pPb Drell-Yan production pseudo-data.

Therefore, assuming full compatibility with the collinear framework, this measurement would provide useful information to improve our knowledge of the gluon density in lead nuclei.

3.2. Constraints from inclusive hadron production

In this case our observable is the cross section of charged hadron single inclusive production in proton-lead collisions divided by the same quantity in proton-proton collisions. Two scenarios were considered: one with pseudo-data generated from the collinear DGLAP framework and EPS09 and another using CGC predictions. The latter is of particular interest to study the discriminating power of this measurement regarding saturation dynamics. In both cases the statistical uncertainty was determined from the expected number of events in each data bin, and we have assumed 5% and 7% (uncorrelated) systematic and normalization uncertainties respectively. Theoretical predictions have been computed using the code for NLO inclusive hadron production of [18], modified to account for nuclear effects as discussed in [19].

In figure 3 we show the comparison between the pseudo-data and the EPS09 predictions, for two rapidities: $\eta = 0$ and $\eta = 2$. The re-weighting produces a shift of the central values (suppression for $p_T < 8\ GeV$ and enhancement above said value) and a narrowing of the uncertainties. Then in table 2 we present the values of χ^2/n, $\langle\chi^2\rangle/n$ and N_{eff}. Unsurprisingly, as data and theory are compatible by construction, many replicas remain. The behaviour for both rapidities is similar and thus we expect a compatible modification of the nuclear PDFs.

As happened for DY, the valence is almost unaffected and the sea density decreases with the error band slightly reduced for $x < 10^{-2}$ (plots not shown). In figure 4 we present the modification of the gluon density due to the re-weighting. The central values show an enhancement for $0.07 < x < 0.2$ while for $x < 10^{-2}$ they are suppressed and the error band shrinks about 50%. The displacements of the central curves, less pronounced in the forward region, are fully compatible within uncertainties, and compatible also with the variation from

Journal of Physics: Conference Series **589** (2015) 012021 doi:10.1088/1742-6596/589/1/012021

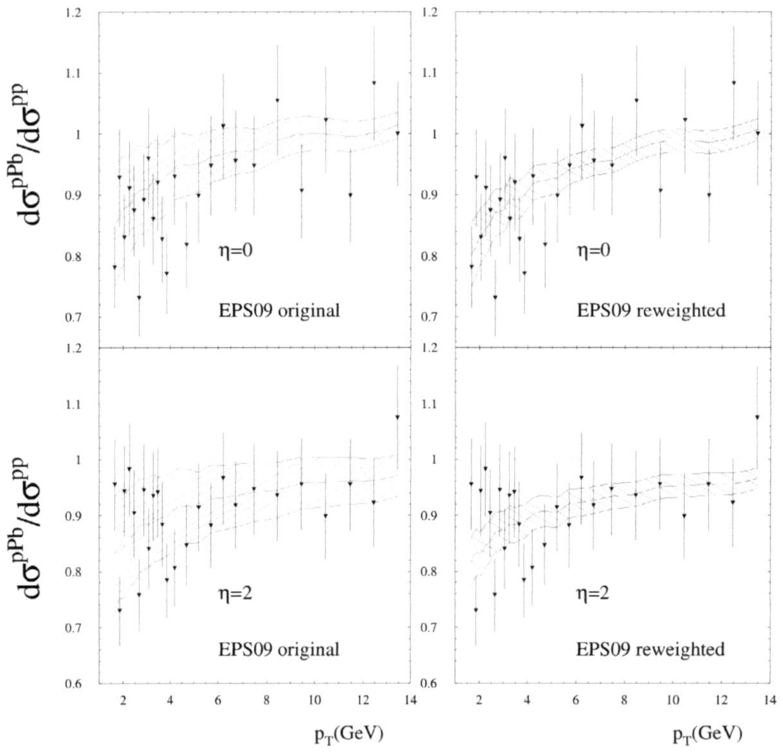

Figure 3. Upper plots: ratio of the charged hadron single inclusive cross section in pPb with respect to pp collisions, as a function of the hadron transverse momentum p_T at central rapidity before (left) and after (right) the reweighting, with pseudo-data generated in the collinear DGLAP framework using EPS09. Lower plots: the same for forward rapidities, $\eta = 2$.

Table 2. Same as Table 1 for the re-weighting of EPS09 with inclusive charged hadron production data, for central ($\eta = 0$) and forward ($\eta = 2$) rapidities. Pseudo-data has been generated in the DGLAP framework.

	$\eta = 0$			$\eta = 2$		
	χ^2/n	$\langle\chi^2\rangle/n$	N_{eff}	χ^2/n	$\langle\chi^2\rangle/n$	N_{eff}
Before	1.11	1.75		0.95	1.82	
After	0.84	1.02	624	0.92	1.08	612

figure 2. Such error reduction in the small-x nuclear gluon ratio is desirable and supports the importance of including this observable in the nuclear PDF fits.

Now we turn our attention to pseudo-data that has been generated in the Color Glass Condensate scenario [12, 20]. In this case the underlying theory for pseudo-data and the one used for the EPS09 predictions are independent, allowing us to analyze the potential discriminating power of this observable. As before, we consider the central and forward rapidity regions. The latter is particularly interesting as data from RHIC in the forward region has not been included so far in nuclear PDFs fits.

We present in figure 5 the nuclear cross section ratios corresponding to $\eta = 0$ (upper plots) and $\eta = 2$ (lower plots) before and after the re-weighting, as a function of the transverse momentum p_T of the final hadron. The corresponding values of χ^2/n, $\langle\chi^2\rangle$ and N_{eff} are shown in table 3.

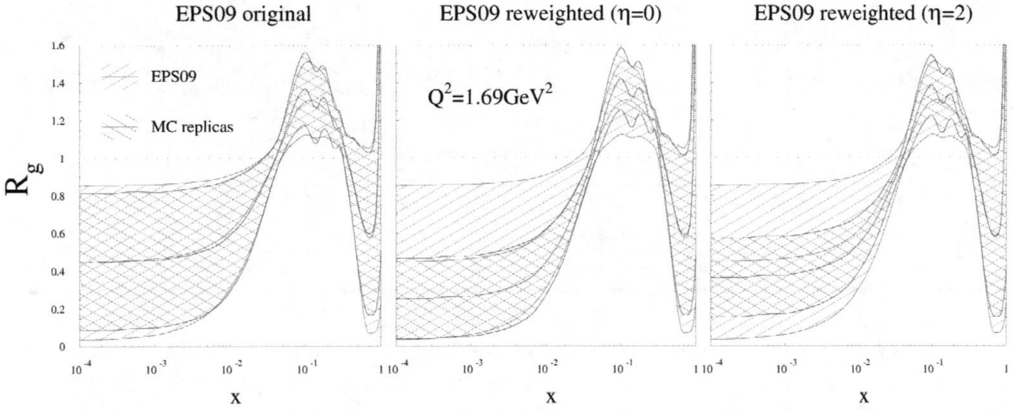

Figure 4. Same as fig. 2, when pPb LHC pseudo-data for inclusive charged hadron production at central ($\eta = 0$) and forward ($\eta = 2$) rapidities are included in EPS09. The pseudo-data has been generated in the collinear DGLAP framework.

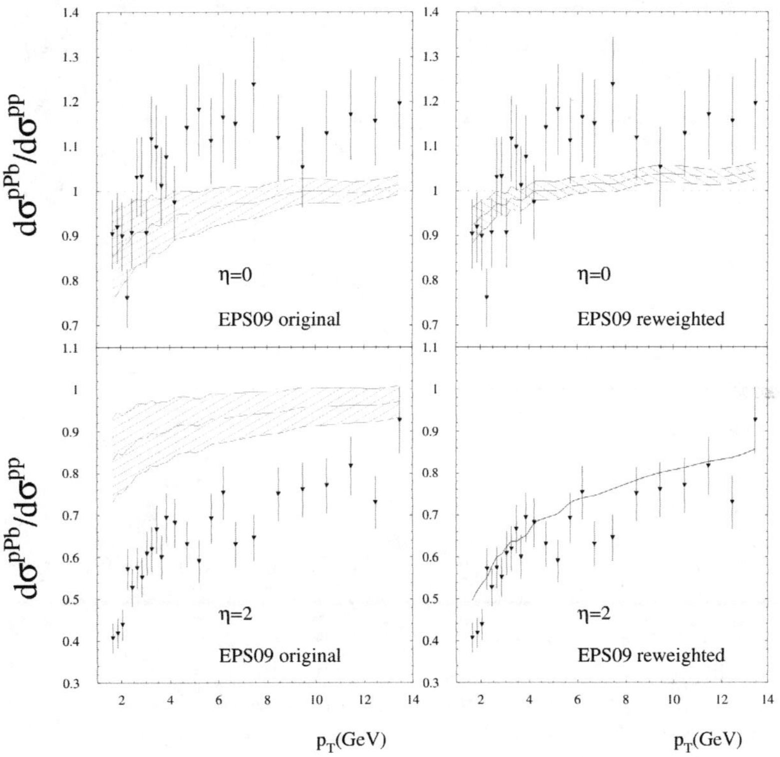

Figure 5. Upper plots: ratio of the charged hadron single inclusive cross section in pPb with respect to pp at central rapidity, as a function of the final hadron transverse momentum p_T before (left) and after (right) the reweighting. Lower plots: the same for $\eta = 2$. Note that the pseudo-data has been generated from the CGC predictions, while EPS09 predictions are based on the collinear DGLAP framework.

Let us begin by discussing the central rapidity region, $\eta = 0$. After re-weighting the agreement between theory and pseudo-data is reasonable. The central value is shifted upwards and the nPDF uncertainties are reduced, leaving 229 effective replicas. The final χ^2/n is above one, pointing the existence of some tension between the CGC and the DGLAP predictions.

Journal of Physics: Conference Series **589** (2015) 012021 doi:10.1088/1742-6596/589/1/012021

Table 3. Same as Table 2 for the case in which pseudo-data has been generated in the CGC framework.

	$\eta = 0$			$\eta = 2$		
	χ^2/n	$\langle\chi^2\rangle/n$	$N_{\rm eff}$	χ^2/n	$\langle\chi^2\rangle/n$	$N_{\rm eff}$
Before	2.25	2.76		36.43	38.62	
After	1.50	1.58	229	1.85	1.85	1

Regarding the impact on the nuclear PDF ratios, the behaviour is different from the one seen with DGLAP pseudo-data: the valence distributions are affected, presenting a non-negligible decrease of the central value but no noticeable change is seen for the sea densities (plots not shown). As for the gluons (figure 6) the central value moves upwards and the uncertainty shrinks around 30% for $x < 10^{-2}$. This trend is opposite to the one seen in figure 4, but they are not incompatible and thus data from the central region is not likely to have the power to discriminate between the two production scenarios.

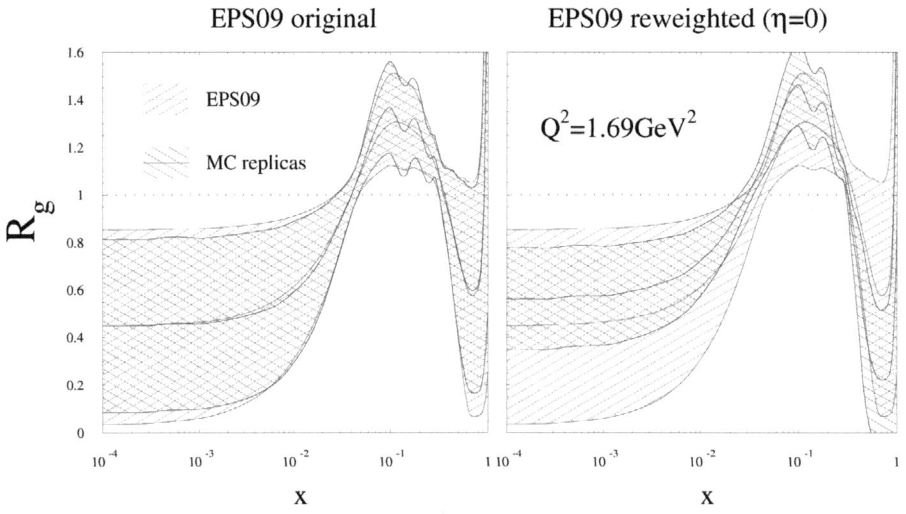

Figure 6. Same as fig. 2, now for pseudo-data generated in the CGC framework.

Now we move to the forward region. The inclusion of the, in principle, inconsistent data is not impossible (lower right plot in figure 5), since the re-weighting brings down the original χ^2/n from 36.4 to 1.8 (see table 3). However, the remaining number of replicas is only $N_{eff} = 1$, implying that almost all MC replicas have been discarded, and just retaining the one that gives the best agreement with the CGC pseudo-data. This is a situation of extreme incompatibility and the re-weighting method breaks down, becoming unreliable. In turn, our result suggests that the differences in forward charged hadron production between the DGLAP and CGC frameworks are substantial and that the LHC has the potential for the discrimination between the two scenarios.

4. Summary

We have presented a study of the potential of LHC proton-lead measurements to constrain nuclear PDFs, based on simulated data. We applied the technique of Bayesian PDF re-weighting and considered two representative processes sensitive to gluons and sea quarks: Drell-Yan production and inclusive charged particle production. In the first case we found that the data has the potential to reduce the PDF uncertainties on the small-x nuclear sea quarks and specially in the medium and small-x nuclear gluon (up to a factor two). In the second case we analyzed one set of pseudo-data generated using the CGC scenario and another one generated using the collinear DGLAP framework. In the latter we found a similar behaviour as for Drell-Yan pseudo-data. However, when pseudo-data comes from CGC predictions the global nuclear fit can be adapted to absorb non-linear effects only in the central rapidity region. In the forward rapidity region, the manifest incompatibility between data and theory breaks down the re-weighting. Regardless the behaviour that real data might have, this process is interesting: constraints on virtually unknown nuclear PDFs can be derived if the non-linear effects are small and, if they are large, experimental accuracy could be enough to identify the onset of the saturation dynamics.

Then our study, based on simplified assumptions, gives a first estimate of the potential of the proton-lead data to constrain nPDFs. Of course when the actual LHC measurements become available we will be able to quantify the impact of the data on the nuclear PDFs. We also release a public version of the Monte Carlo EPS09 replicas, that allows the experimental groups to study the impact of their data on nuclear PDFs, without the need to wait for an updated fit. The driver code, data files and documentation can be obtained from **http://igfae.usc.es/hotlhc/index.php/software**.

Acknowledgments

I thank my collaborators on the work described here: N. Armesto, J. Rojo and C. A. Salgado. J. R. was partially supported by a Marie Curie IntraEuropean fellowship of the European Community's 7th framework Programme under contract number PIEF-GA-2010-272515. This work is supported by European Research Council grant HotLHC ERC-2011-StG-279579; by Ministerio de Ciencia e Innovación of Spain under projects FPA2009-11951 and FPA2011-22776, and the Consolider-Ingenio 2010 Programme CPAN (CSD2007-00042); by Xunta de Galicia (GRC2013-024); and by FEDER. I thank Kari Eskola and Hannu Paukkunen for useful discussions, and Javier Albacete for providing us with the CGC predictions. I am also grateful for the hospitality of the CERN TH Unit where part of this work was performed.

References

[1] Salgado C A, Alvarez-Muniz J, Arleo F, Armesto N, Botje M, Cacciari M, Campbell J and Carli C *et al.* 2012 *J. Phys. G* **39** 015010
[2] Albacete J, Armesto N, Baier R, Barnafoldi G G, Barrette J, De S, Deng W-T and Dumitru A *et al.* 2013 *Int. J. Mod. Phys.* E **22** 1330007
[3] Paukkunen H and Salgado C A 2011 *JHEP* **1103** 071
[4] Stavreva T, Schienbein I, Arleo F, Kovarik K, Olness F, Yu J Y and Owens J F 2011 *JHEP* **1101** 152
[5] Arleo F, Eskola K J, Paukkunen H and Salgado C A 2011 *JHEP* **1104** 055
[6] Kang Z B and Qiu J W 2013 *Phys. Lett.* B **721** 277
[7] Eskola K J, Paukkunen H and Salgado C A 2013 A perturbative QCD study of dijets in p+Pb collisions at the LHC (*Preprint* hep-ph/1308.6733)
[8] Ball R D *et al.* [NNPDF Collaboration] 2011 *Nucl. Phys.* B **849** 112, 2012 [Erratum-ibid. B **854** (2012) 926], [Erratum-ibid. B **855** (2012) 927]
[9] Ball R D, Bertone V, Cerutti F, Del Debbio L, Forte S, Guffanti A, Hartland N P and Latorre J I *et al.* [NNPDF Collaboration] 2012 *Nucl. Phys.* B **855** 608
[10] Armesto N, Rojo J, Salgado C A and Zurita P 2013 *JHEP* **1311** 015
[11] Eskola K J, Paukkunen H and Salgado C A 2009 *JHEP* **0904** 065
[12] Armesto N 2006 *J. Phys.* G **32** R367
[13] Ball R D *et al.* [NNPDF Collaboration] 2009 *Nucl. Phys.* B **809** 1, [Erratum-ibid. B **816** 293]
[14] Ball R D, Del Debbio L, Forte S, Guffanti A, Latorre J I, Rojo J and Ubiali M [NNPDF Collaboration] 2010 *Nucl. Phys.* B **838** 136
[15] Watt G and Thorne R S 2012 *JHEP* **1208** 052
[16] Golec-Biernat K, Lewandowska E and Stasto A M 2010 *Phys. Rev.* D **82** 094010
[17] Campbell J M and Ellis R K 2000 *Phys. Rev.* D **62** 114012
[18] Jager B, Schafer A, Stratmann M and Vogelsang W 2003 *Phys. Rev.* D **67** 054005
[19] Sassot R, Stratmann M and Zurita P 2010 *Phys. Rev.* D **82** 074011
[20] Albacete J, Dumitru A, Fujii H and Nara Y 2013 *Nucl. Phys.* A **897** 1